Satellite Network Robust QoS-aware Routing

Fei Long

Satellite Network Robust QoS-aware Routing

Fei Long
Beijing R&D Center of 54th Research
Institute of China Electronics Technology
Group Corporation
Beijing
China

ISBN 978-3-662-52426-8 ISBN 978-3-642-54353-1 (eBook)
DOI 10.1007/978-3-642-54353-1
Springer Heidelberg New York Dordrecht London

Jointly published with National Defense Industry Press, Beijing

ISBN: 978-7-118-09491-6 National Defense Industry Press, Beijing

Printed on acid-free paper

Springer is part of Springer Science+Business Media (www.springer.com)

Preface

The satellite network technology is derived from the technology of satellite communication of the 1950s. The initial mode of satellite communication is bent pipe, which sends back to earth what goes into the conduit with only amplification and a shift from uplink to downlink frequency. The satellite network has come true after the emergence of on-board processing and inter-satellite links. The Iridium constellation is a typical example of satellite network. The satellite network developed rapidly since the concept was proposed, and has attracted more and more attention in the recent years. Compared with terrestrial network, satellite network has larger coverage range, easier accessory, and hence is widely used in environmental monitoring, navigation, and resource exploration. Moreover, satellite network can provide wireless access to the Internet for residents in remote areas.

As a core technology of the satellite network system, satellite network routing plays an important role in determining the quality of service of the whole system. Since the environment of the satellite network is quite different from the terrestrial network, the routing of the terrestrial network cannot be used directly in the satellite network. This book analyzes the characteristics of the satellite network, and proposes a series of robust satellite network routing methods including topology design, routing scheme, traffic engineering, and QoS routing algorithms.

The book consists of six chapters. Chapter 1 introduces the basic concept of the satellite network, and the current research status of the satellite network routing. The technical difficulties of satellite network routing are also presented in this chapter. Chapter 2 presents a set of constellation design and optimization schemes. A constellation model that satisfies global coverage, low layer coverage, low system cost, and long continuous coverage time is proposed. The influence of the constellation structure on the network transmission performance is explored through simulations. A novel satellite network virtual grouping routing strategy is proposed in Chap. 3. The dividing method of time slots is improved, the congestion avoidance and node failure handling are added in this chapter, which promotes the robustness of the satellite network routing strategy. A satellite network traffic prediction method based on time series analysis theory according to the terrestrial network traffic information and global user distribution information is proposed in Chap. 4. A traffic engineering method that uses the traffic prediction as the input and maximum link utility as the optimization objective is also presented in this chapter, which greatly increases the robustness of the satellite network to the unpredicted traffic spike. Chapter 5 proposes a QoS routing algorithm that fits for the environment of satellite network. Heuristic algorithms such as ant colony, genetic algorithm as well as some multiobjective optimization algorithms such as PEC and prior-order algorithm are introduced to optimize for different QoS combination of user requirements, which solves the problem of multi-QoS-objective optimization that cannot be solved by traditional SPF routing algorithm. Finally, the summary and outlook of the research content of this book are made in Chap. 6.

The author has worked for nearly 10 years in research on satellite network routing, and joined the National High Technology Research and Development Program of China (863 Program) and the National Natural Science Foundation of China programs on satellite network routing. This book is a summary of his research experience in constellation design, satellite network routing scheme, and satellite robust QoS routing. We hope that this book will provide useful reference for researchers in the field of satellite network routing, and believe that the publication of this book will promote the theoretical research of the spatial information system.

Beijing, China, June 2013

Wu Manqing
Yu Quan

Contents

Chapter 1
Introduction

1.1 Background

With the rapid development of globalization, the Internet has become an ndispensible part of people's daily lives. The traditional Internet mainly depends on terrestrial link transmissions, which can hardly cover complicated terrains, such as highlands and islands. Terrestrial link disruptions caused by earthquakes or road digging happen almost every year. This causes huge economic losses to society. For example, two backbone links (Oroville-Stockton, Rialto-El Palso) of a major US Internet service provider were cut due to a road digging error, which disconnected millions of Internet users in the western United States from the Internet for a few hours. The damage was estimated at several million dollars [1]. Other examples are the September 21, 2007, earthquake in Taiwan and the May 12, 2008, Wenchuan earthquake in China's Sichuan Province. The local Internet systems suffered a devastating blow in both of the two earthquakes.

The above-mentioned shortcomings of the traditional terrestrial networks and the growing demand of Internet users make satellite networking gradually become the research focus in many nations. Due to the characteristics of global coverage, easy access, excellent expandability, and variable bandwidths on demand, satellite networks are widely used in weather forecasting, environmental and disaster monitoring, resource explorations, navigation, positioning, and broadcasting communications, etc. Undoubtedly, it will become an important part of the next generation Internet (NGI) [2]. Therefore, the level of space information technology presented by satellite networking is becoming one of the symbols of the comprehensive national strength.

Since the space satellite orbit resources are very limited, not only have the various space powers, led by the United States and Russia, begun their research and construction of satellite networks, but more and more small and medium-sized countries have also strengthened their development efforts in space information technology. Currently, more than 130 countries in the world have carried out research and development of spatial information technology. It is expected that the whole world will invest over $ 500 billion in developing and launching more than

F. Long, *Satellite Network Robust QoS-aware Routing*,
DOI: 10.1007/978-3-642-54353-1_1,

1,000 various types of satellites. China attaches great importance to the development of aerospace science and space information technology to adapt to the trend of the world's technological development. In order to shorten the gap with the leading countries on a technical level, the Ministry of Science and Technology of China and the Chinese Academy of Sciences take the aerospace integrated information network as an important research direction in the country's 11th Five-Year Development Plan. Since the launch of its first self-developed satellite, Dongfanghong-1 in 1970, China has successfully launched a total of 63 self-developed satellites of 18 types, widely used in various fields of economy, science and technology, culture, and national defense. At present, China has launched a series of satellites and formed several satellite network systems, such as the "Vanguard" series of retrievable reconnaissance satellites, the "Practice" series of satellites, the "Resource I" earth observing satellite, the "Storm rider" series of meteorological satellites, the "Supernatural powers" strategic communications satellites, the "Vanguard III" military transmission satellite, "Vanguard V" radar satellite, "Beacon I" communications satellite, "Big dipper" navigation and positioning satellite, and "Ocean I" target surveillance satellite etc.

Besides the major applications in meteorology, reconnaissance, and communications and positioning, satellite networking is also an important part of the Interplanetary Internet (IPI) [3]. Scientific explorations of deep space, such as Mars explorations, will produce a large amount of scientific data, and the transmission of these scientific data depends on a high data transfer rate, high security, and seamless interoperability interstellar networks. As the last hop of the deep space data transmission, satellite networks are responsible for data relay, data fusion, and command transmission etc. It is also an important research field in NASA's deep space network (DSN) operations [4, 5].

Since the concept of satellite network was proposed, it has been mainly in use as a bypass network of terrestrial networks and provides services for real-time or non-real-time applications. Satellite network routing technology as a key part of the satellite network technology determines the efficiency and reliability of the entire satellite network system. Due to the cyclical changes in the high-speed of satellite network topology, the traditional terrestrial network routing protocols, such as OSPF [6] and RIP [7], cannot be directly applied to satellite networks.

OSPF and RIP protocols are widely used for fixed terrestrial network topology. Each time when the network topology is changed such as when a link is established or fails, or when a node joins the network or becomes invalid, OSPF and RIP will re-exchange the network topology information and refresh the routing table. The high-speed mobility of a satellite node in a satellite network will lead to a frequent change of network topology information, and cause great overheads to OSPF and RIP protocols.

Although an ad hoc network also has the topology time-varying features, its routing protocol cannot be used in satellite networks either. This is because the characteristic of mobility is quite different between the two networks. First, the topology of a satellite network changes all the time while the topology of ad hoc network changes only at certain times, Second, the satellite network topology changes at high speed. Take a LEO (Low Earth Orbit, LEO) satellite with an orbital altitude of 500–2000 km as an example, the duration when it is visible at a fixed point on the ground is only about 10 min, while the nodes in an ad hoc network moves much slower. Lastly, an ad hoc network is self organized, and its topology change is not regular. In contrast, the topology change of a satellite network has a strictly periodic law. Therefore, the existing ad hoc network protocols, such as DSDV, TORA, AODV, and DSR [8], cannot be applied to satellite networks either.

Compared with terrestrial networks, satellite networks have far fewer backup routers and more complicated topology, and hence are more in need of robust routing algorithms. As an important complement to terrestrial networks, satellite networks need to provide reliable network transmission services for different types of users, which make a great demand on the design of QoS routing protocols. The topological characteristics of satellite networks, network environment characteristics, and the quality of service on the satellite network expectations make the satellite network's robust QoS routing protocol designs face the following challenges:

(1) The low overhead and high reliability requirements of the protocol. The energy supply and computing and storage resources of a satellite network system are very limited. Moreover, the maintenance of the onboard equipment is inconvenient. In order to improve the system's life cycle, the routing protocol is required to have low overhead and high operational stability.
(2) The contradictions between the user QoS requirements and system performance. The large propagation delay and the high error rate of satellite links conflict with the low-latency and high-accuracy transmission requirement of users. In addition, QoS requirements of different users are not identical, thus requiring the routing protocol to take certain strategies to reduce the error rate, prevent congestion, and shorten the transmission delay, while satisfying the QoS requirements of different users simultaneously.
(3) The robustness requirement of a satellite network system. As a bypass network of terrestrial networks, satellite networks carry a similar type of network traffic as terrestrial networks. The robust routing problem of terrestrial networks also exists on satellite networks, and the limitedness of resources of satellite networks makes robust routing for satellite networks more challenging. There exist unpredictable traffic spikes on the network, and the network traffic at different times is not the same. This requires a robust routing protocol which can make the system work well under different network traffic flows, even when traffic spikes occur unpredictably.

In order to solve the above-mentioned problems in satellite network routing protocols, this book intends to explore a QoS routing method, which can provide QoS transmission services for different users under resource-constrained conditions, and ensures stable performance under complex network traffic flows. It is the author's sincere hope that the ideas offered in this book may shed some light on future research on the subject.

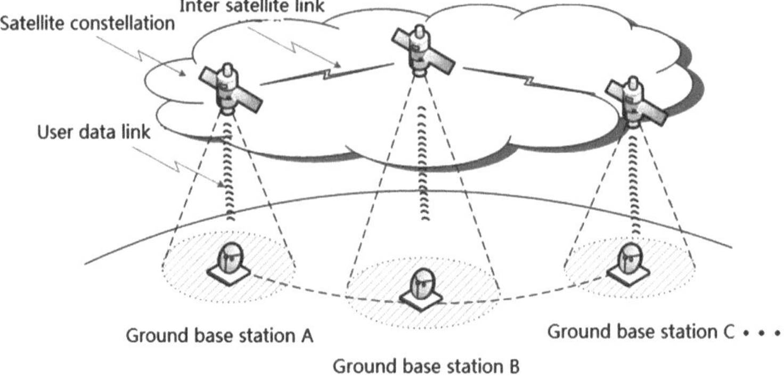

Fig. 1.1 Sketch map of satellite network

1.2 Introduction to Satellite Networking

A satellite network system is usually composed of three parts: a ground base station, the satellite constellation system, and up/down links as shown in Fig. 1.1. Ground base station includes a gateway station and control center. The control center is responsible for the operation and management of the entire satellite network resources, satellite remote operation, and orbit control. The gateway station provides the interface between different external networks and satellite networks, and it is also in charge of protocol conversion and address translation [9]. A satellite constellation system consists of different satellites in different orbits connected by intersatellite links. Satellite nodes in a constellation system have the ability of onboard processing. They play the same role as ground routers and are responsible for user packet routing and forwarding.

Satellite constellations fall into three categories based on the altitude of a satellite orbit: Geosynchronous Earth Orbit (GEO) satellite constellations, Medium Earth Orbit (MEO) satellite constellations, and Low Earth Orbit (LEO) satellite constellations. They can also be divided into circular orbit satellite constellations and elliptical orbit satellite constellations according to the shape of the orbit. A focal point of the elliptical satellite orbit coincides with the center of the earth while the center of the circular satellite orbit coincides with the center of the earth, too. GEO, MEO, and LEO satellites all belong to the circular orbit satellites. Their relative positions to the Earth are shown in Fig. 1.2.

Satellite constellations can be divided into polar orbit constellations and inclined orbit constellations according to their orbital inclinations. They can also be divided into single-layered constellations and multilayered constellations according to the number of satellite layers.

A GEO satellite's orbital plane coincides with the Earth's equatorial plane, and the altitude of the orbit is 35,786 km. In this orbit, the angular velocity of the satellite is the same as the speed of the earth's self-rotation. This makes the

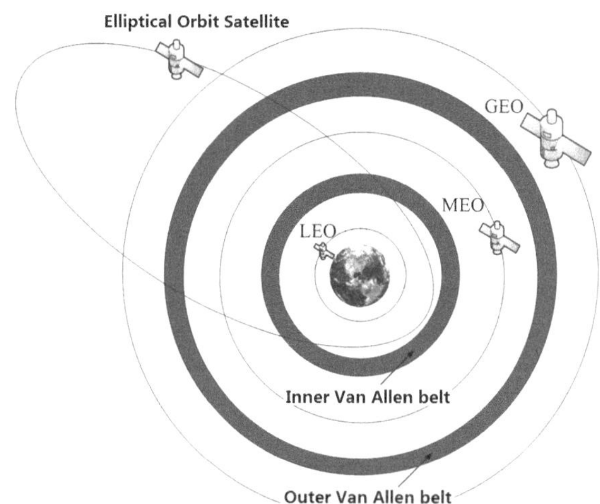

Fig. 1.2 Type of satellite orbit

satellite seem stationary when observed from a fixed point on the ground. As a result, a GEO satellite is also called a geostationary orbit satellite. The interval between two adjacent GEO satellites is determined by the minimum elevation angle of the antenna on the ground. Usually the angle can be as small as 1.5° [10]. The service area of a GEO satellite is vast. Theoretically, three GEO satellites can cover the whole earth except the Polar Regions. The propagation delay between the ground base station and a GEO satellite is related to the latitude and longitude of the station, usually between 125 and 250 ms, which is often referred to as a half-second round-trip delay (Round-Trip Time, RTT). GEO satellites are commonly used in the bent-pipe relay network, which has no onboard processing ability, such as the ultra small earth station satellite communications system (Very Small Aperture Terminal, VSAT) [11].

The altitude of MEO orbits ranges between 9,000 and 11,000 km, which are located between the inner and outer Van Allen belts. The Van Allen belts are two radiation belts outside the Earth's atmosphere, located, respectively, 2,000–5,000 km (proton belt) and 13,000–19, 000 km (electronic belt) from the Earth. The charged particles in the two radiation belts can cause large interferences on the satellite electronic equipment, so an MEO satellite orbit should be kept away from these two belts. The visible time of an MEO satellite for a fixed point on the ground lasts for a few dozen minutes with an average RTT of 110–130 ms [1]. The ICO (Intermediate Circular Orbit) system is an example of application for a MEO satellite constellation [12].

The altitude of LEO orbits ranges between 500 and 2,000 km, which are located between the inner Van Allen belt and the upper atmosphere. LEO satellites revolve around the Earth at a high speed of about 25,000 km/h, and the visible time for a fixed point on the ground is about 10 min with an average RTT of less than 30 ms. The low RTT value of LEO constellations makes it very suitable for some time-sensitive communications. The application examples of LEO

Constellations include the Iridium Constellation of Motorola [13], the Teledesic Constellation [14], and the Globalstar Constellation of America's LQSS (Loral Qualcomm Satellite Service) company[15].

The orbital velocity of an elliptical orbit satellite changes as its position changes. It is slow at apogee and fast at perigee. Since an elliptical orbit satellite can serve for the ground only when it nears its apogee when its velocity is slow, and its coverage area is limited to high-latitude regions at that time, only a small number of high-latitude countries, such as Russia, use elliptical orbit satellites. Generally, the included angle between the elliptical orbit satellite's orbital plane and the equatorial plane is 63.4°, for this type of satellite stays relatively stationary with the ground. The Molniya system and the Tundra system [16] of Russia are two typical elliptical orbit satellite systems. Due to the limited application of elliptical orbit satellites, they are not discussed in this book.

The included angle between the satellite orbit plane and the equatorial plane is referred to as an orbital inclination angle. The satellites with a 90° orbital inclination angle will pass over the Earth's north and south poles; therefore, they are also called polar orbiting satellites. The constellation that is formed completely with polar orbiting satellites is called a polar orbiting constellation. The polar orbits are frequently used by Earth mapping satellites, remote sensing satellites, reconnaissance satellites, and some meteorological satellites. The Iridium constellation mentioned above is a polar orbiting constellation. The satellites with their orbital inclination angles being different than 90° are called inclined orbit satellites. The constellation that is formed completely with inclined orbit satellites is called an inclined orbit constellation. A typical inclined orbit constellation is the ICO system, which is composed of 10 inclined orbit satellites evenly distributed in two 45° inclined orbits.

A satellite constellation can be composed of single-layered satellites and multilayered satellites. A single–layered constellation can be composed of LEO satellites, or MEO satellites, or GEO satellites. A multilayered constellation can be composed of LEO, MEO, and GEO satellites with orbits being an arbitrary combination of any two of the three or just of the three. For example, the Spaceway system built by Hughes Communications Galaxy Inc., [16] is a multilayered constellation. Its coverage regions include North America, Latin America, Asia-Pacific, Europe, Africa, and the Middle East. Each area is served by GEO and MEO satellites together. Satellites in the same layer are connected by Inter Satellite Links (ISL), and satellites in different layers are connected by Inter Orbit Links (IOL).

The satellite–ground links that connect the ground base station and the satellite constellation are also known as User Data Links (UDL). An interface satellite can connect with several ground base stations, while a ground base station can also connect with several interface satellites. The bandwidth of UDL is usually 2 Mbps [17].

1.3 Current Status of Research on Satellite Network Routing

The purpose of satellite network routing research is to develop and design efficient routing algorithms and protocols according to the characteristics of certain satellite networks to provide a reliable data transmission path for the users. Early satellite network systems used a single GEO satellite to relay data in a bent-pipe mode, not involving any routing technology. In a satellite network consisting of multiple satellites, an optimal path is needed to be selected from multiple available source–destination paths according to link metrics or user QoS requirements. Satellite network routing technology plays an important role in determining the efficiency and reliability of the entire network system. According to the structure of a satellite network system, satellite network routing can be divided into three parts.

(1) Border routing: A satellite network is deemed as an important part of NGI. A satellite network and terrestrial network border routing protocol needs to be developed for the integration of the terrestrial network and the satellite network. The protocol can achieve seamless interoperability between terrestrial networks and satellite networks. The satellite–terrestrial routing protocol should be run at the ground base station of each ground autonomous system (Autonomous System, AS). It is responsible for the discovering and broadcasting paths throughout the satellite network, playing the same role as the Border Gateway Protocol (BGP) in terrestrial networks.
(2) Access routing: Also known as UDL routing, it is responsible for finding an access satellite for the ground mobile users and ground base stations generally according to the UDL survival time, delay, and signal strength.
(3) Intersatellite routing: When data packets are uploaded to an access satellite, intersatellite routing is responsible for finding one or several paths from the access satellite to the export satellite to satisfy certain QoS requirements.

1.3.1 Border Routing

Problems with border routing occurred from the integration of satellite networks and terrestrial networks first proposed in [19–21]. These papers propose that the satellite network system can be viewed as an independent AS system with different address mechanisms. To alleviate the burden of the satellite network, the terrestrial gateway should be used as the border gateway of the satellite network, responsible for the translation of the network address. Therefore, the external gateway protocols of a traditional terrestrial network, such as the BGP protocol, can be used to route between the satellite network and the terrestrial network [18]. The integration of satellite and terrestrial networks is difficult due to the different internal and external parameters, and the above-mentioned papers do not offer any specific solutions. A satellite version of the Border Gateway Protocol (BGP-S) was first proposed in [19] and improved in [20]. In these two papers, a satellite network is deemed as an AS system with special attributes, and BGP-S is only implemented in the terrestrial gateway connected to the satellite network. It is compatible with BGP-4 and can support the auto discovery of paths including satellite nodes. The design and implementation of BGP-S has been elaborated in [2].

1.3.2 Access Routing

Access routing is relatively simple to implement. The mission of an access router is to find an interface satellite for ground mobile users or terrestrial base stations according to the coverage time of the satellite and the link signal strength of UDL. The LEO satellite which has the longest coverage time is selected as the interface satellite as suggested in [21] the advantages of doing so is to reduce the UDL link switching and protocol overheads. While in [22], the LEO satellite is chosen as the interface satellite for the ground base station according to the signal strength of UDL. The satellite having the strongest UDL signal usually is the satellite nearest to the ground base station and has the largest elevation from the base station. The calculation is simple and the current communication quality can be guaranteed in this selection method.

1.3.3 Inter Satellite Routing

Intersatellite routing is the key and most difficult part of the entire satellite network routing technology. The bulk of the current research on satellite network routing technology focuses on this part. In order to find the path that meets certain requirements from the source satellite to the target satellite, the time-varying topology of the satellite network must be solved first. The strategy to solve the problem of satellite network topology change for the purpose of routing is called satellite network routing strategy [23]. The satellite network routing strategy is the basis for the application of the satellite network routing algorithm. Satellite network routing strategies can be divided into two kinds: the virtual node strategy and the virtual topology strategy.

The virtual node strategy was first proposed by Ekici E in [24]. According to this strategy, Satellite S is identified by Longitude S and Latitude S. It is assumed that the whole earth is divided into different logical areas based on different latitudes and longitudes. Each logical area is bound with the nearest satellite. When a LEO satellite (i, j) leaves its current logical area at the moment t, the next LEO satellite $(i, j - \Delta)$ that enters this logical area will take over (i, j)'s position. Satellite (i, j) will transfer its routing table and all the information associated with the current logic area to the satellite $(i, j - \Delta)$. Satellite $(i, j - \Delta)$ is thus called the

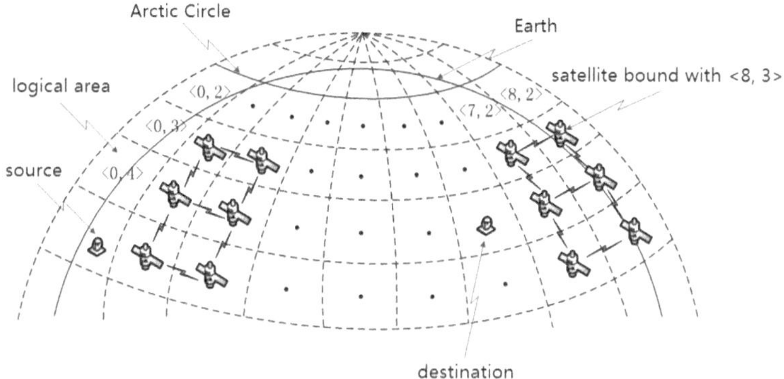

Fig. 1.3 The principle of virtual node strategy

successor of satellite (i, j). Therefore, the ground logical area can be identified as $<p, s>$ $p = 0,\ldots, N - 1$; $s = 0,\ldots, M - 1$, where N and M represent the numbers of orbits in the LEO layer and the number of satellites in each orbit, respectively. As shown in Fig. 1.3, the virtual node topology shields the time-varying topology of the satellite network. This means that, only a fixed ground logical area is needed to be considered in calculating a path from the source ground base station to the destination ground base station, and the mobility of the satellite nodes does not need to be taken into account in the routing calculation. Although the virtual node topology proposed in [24] is based on the LEO layer satellite network, it can be extended to any single-layered satellite networks in practice.

The virtual topology strategy was proposed by Werner M in [25]. The basic idea of this strategy is to divide the constellation period of revolution into several periods of time according to the predictability and periodicity of the constellation's movement. The constellation topology in each period of time can be deemed as fixed. The fixed period of time is hence referred to as a topology snapshot or time slot. The routing algorithm can calculate the route in each time slot.

A routing framework based on the coverage domain division proposed in [26] is considered to be the prototype of the virtual node strategy. The framework utilizes the thought of mobile IP and hierarchical IP addressing, and divides the Earth's surface into a certain number of cells and macro cells according to the coverage area of the satellite. A satellite node finds the next hop satellite for the packets according to the cell and macro cell ID until it reaches the destination interface satellite. The paper uses the Teledesic constellation as a model and divides the Earth's surface into several macro cells with a side length of 160 km. Each macro cell is in turn divided into 9 cells. The cellular area is marked with (MID, CID), where MID is its macro cell number and CID is its cell number. The binding criterion is the receiving power, that is to say, the nearest satellite will be bound with the cell. The packet header is defined as <MID, CID and Terminal ID, Datagram ID, Sequence No., TTL>, where the Terminal ID marks the terminal of the destination. MID, CID, and Terminal ID are related to the routing. These three parts compose the virtual address of the destination, and can invariably mark a destination port. User packets are split at the source terrestrial gateway into satellite network packets, and then forwarded to the interface satellite bound with the cell. Since each satellite node stores the geographic location of the entire constellation, packets entering the satellite network can be forwarded to the export satellite and down to the destination ground gateway through UDL. Work in [27] is similar to that of [26]. Based on [26], the distributed geographic routing algorithm (DGRA) is proposed in [27]. Different from [26], DGRA forward packets in two steps. When the distance between the packet and the export satellite is greater than a certain threshold, the satellite forward the packet according to the global routing information, otherwise, it will forward the packet based on the local routing information.

The routing framework proposed in [26, 27] based on coverage area division basically has a virtual node routing policy thinking. It is relatively simple to implement. However, it lacks the processing of link congestion and the temporarily satellite link shutdown in the Polar Regions. As a result, it is poor in robustness. The datagram routing algorithm (DRA) proposed by Ekici E in [24, 28] is a typical virtual node routing algorithm. DRA optimizes the end-to-end delay which is the same as the Shortest Path First (SPF) algorithm. In DRA, the satellite network is deemed as a mesh system that contains the logical location information. Packets are routed in this fixed topology in a distributed manner. The so-called logical location information is a 2-tuple $<P, S>$ that identifies the spatial location of an LEO satellite, where P is the orbit number of the satellite and S is the satellite number in the orbit. Since the algorithm is applied to a polar constellation, P is constant in the routing computation and S changes according to the bound terrestrial area.

DRA conducts two-stage routing based on the SPF algorithm. In the first stage, the satellite that receives the packet first assumes that the weights of all ISLs in the network are equal, and then computes the route for the packet according to the satellite's own $<P, S>$ value, the $<P, S>$ value of the destination satellite, and the constellation topology information it has now. Since the topology of a polar satellite constellation is symmetrical and the weights of ISLs are the same, the computational results may not be invariable. And then in the second stage, the algorithm will choose an optimized route according to the total ESL length from the available routes obtained in the first stage. In the second phase of routing, the inner orbit ISL has a higher priority to be elected than the interorbit ISL with the same length. This is because the inner orbit ISL is permanent without link switching. But the interorbit ISL is a temporary link, and will be interrupted when entering the polar region or when the two satellites cannot see each other. DRA has a congestion processing function by monitoring the link export queue buffer occupancy to detect and avoid congestion. When the buffer queue reaches a threshold, it will send back a warning signal to notify the source satellite to reroute in order to avoid congestion.

DRA proposed the virtual node routing strategy explicitly for the first time, and utilized the symmetry of a polar orbiting constellation for routing. The implementation of DRA is simple. Since there are no exchanges of any topology information between the satellites, the algorithm does not have any additional communications overhead [2], and the routing selection time is very short. However, precisely because there are no topology information exchanges, each satellite can only get topology information of the entire constellation from the ground base station it binds, which is computed in advance. This makes the performance of the algorithm degrade sharply when a satellite or ISL fails. As a result, the robustness of this algorithm is poor. DRA is applied to a polar LEO constellation, whose topology is relatively simple. So there needs great improvement when it is extended to the inclined orbit constellation or other constellations with more complex topology.

The Probabilistic routing protocol (PRP) is proposed in [29, 30] based on the virtual node strategy. PRP utilizes the predictability of the constellation movement and the statistical probability of the call to route, in order to reduce rerouting due to link switching. In the routing calculation, PRP ignores the ISLs that will be shutdown or switched in a high probability during the call. If the probability is larger than p, the ISL will be removed in the calculation. The switching probability can be calculated according to the probability density function of the effective coverage domain of the LEO satellite. The advantage of PRP is that it can guarantee the stability of the link during a call to a certain probability, hence reduces the protocol overhead caused by link switching. And the disadvantage of PRP is that it depends on the value of p greatly. If the value of p is high, the rerouting probability caused by link switching is low. However, a high value of p also leads to relative concentrated route selection, which will increase the blocking probability of a new call. Unfortunately, there is no quantitative analysis of the relationship among the probability p, the new call blocking probability, and the rerouting probability caused by link switching in [29].

A distributed routing algorithm called the Local Zone-Based Distributed Routing (LZDR) was proposed in [30]. LZDR is also based on the virtual node strategy. The algorithm is slightly different from the DRA algorithm. The routing domain is not divided in accordance with a single virtual node. The algorithm merges a few adjacent nodes into one zone. LZDR conducts routing in two phases: interdomain routing and intradomain routing. LZDR first defines the management node in each zone, and then calculates the best path from the source management node to the destination management node using the hop count as the optimization target.

In the intradomain routing phase, nodes need to exchange status information. The algorithm computes the path from the destination management node to the destination node according to delay. Implementation details of LZDR are not given in [30], such as how to determine the boundaries of the zones. The network status information is restricted in each zone, which makes the traffic demand among zones unknown. That means the interdomain routing may not be optimized and the congestion may occur easily. Although LZDR simplifies the routing computing by modeling the polar orbit constellation as a Manhattan Street Network, MSN [31], the extensibility of the algorithm is not considered.

Besides, algorithms proposed in [32–36] are also based on the virtual node strategy. Some improvements are made in PRP based on virtual nodes in [32], which reduce the link switches while optimizing the end-to-end delay. A routing algorithm based on IP in a multihop satellite network is proposed in [33]. It reduces the link switches in each call while satisfying the QoS requirements of users. An IP routing algorithm in a quasipolar constellation [34] binds the satellites to the logical areas dynamically, and then transplants the terrestrial IP routing technology into the satellite constellation in order to implement the seamless operation between the satellite network and the terrestrial network. Moreover, the satellite constellation routing based on logical topology in [35] is similar to it. The load balance method in an LEO satellite constellation proposed in [36] is also based on the virtual node strategy. It optimizes the maximum link utility (MLU) in the routing computing.

Although the virtual node strategy is simple and the protocol overhead is little, it can only be applied to a single-layered satellite constellation and can hardly be extended. Almost the entire virtual node policy-based routing methods are applied to polar orbit LEO constellations for their regular topology. The virtual node strategy needs to be greatly improved if it is extended to an inclined orbit satellite constellation, which will increase the computational complexity and protocol overhead. The strategy is not applicable to multilayered satellite constellations for the coverage area of different layers is overlapped which makes the logical area unable to be bound with the only satellite constellation.

Corresponding to the virtual node strategy, the virtual topology strategy is another routing strategy to solve the dynamic topology of a satellite network. It was first proposed in [25]. In this paper, the system cycle of a satellite network is equally divided into N snapshots, and the constellation topology is deemed as fixed in each snapshot. As a result, the dynamic routing problem of a satellite network is transformed into a series of static routing problems in the snapshots. The dynamic routing algorithm proposed in [25] is based on ATM, so the purpose of the algorithm is to find a virtual path (VP) from origin to destination (OD) nodes while satisfying some requirements. Since there may exist link switches caused by snapshot switching, the virtual topology strategy needs to take some measures to reduce the rerouting caused by it. The algorithm in [24] adopts a two-phase routing method. First, it finds the set of delay constraint optimal VPs in the snapshot, and then finds a VP from the set that has the least link switches when snapshot switches. The algorithm is an offline algorithm; the optimal VPs are calculated at the base station in advance and uploaded to a satellite node when needed. The satellite node only needs to modify the routing table in the snapshot switching point. Different from the virtual node strategy, the virtual topology strategy can be extended to multilayered satellite constellations. However, the algorithm is vulnerable as most offline satellite routing algorithms. It cannot reduce the overhead caused by link switch rerouting effectively.

A routing algorithm based on finite state automaton (FSA) is proposed in [37, 38]. The algorithm also divides the satellite system cycle into equal length snapshots, and takes link utility as the optimizing object. Each topology state of snapshot is deemed as a FSA state, which transforms the satellite network routing problem into a link assignment optimization problem for OD nodes in FSA. The simulated annealing (SA) algorithm is used in the routing computation. The joint optimal solution of link assignment and routing is computed using an iterative method according to the visibility matrix of satellite and the traffic constraint matrix. Same as the algorithm proposed in [25], the FSA-based algorithm is also an offline routing algorithm, hence are not dynamic adaptive to traffic load and link congestion. Besides, the FSA routing algorithm can only satisfy the link utility requirement. The rerouting problem caused by snapshot switches is not solved in [37, 38] either.

A routing mechanism based on ATM is proposed in [39]. It divides the satellite network system cycle into several snapshots, and uses the SPF algorithm to find the path with the shortest delay for each OD nodes in each snapshot. The idea of dividing the satellite network system cycle unequally was first proposed in it. In other words, the length of each snapshot is not equal. The snapshot switches when a new ISL is established or an old ISL is cut. The algorithm finds k number of paths which meet the bandwidth and delay constraints for each pair of OD nodes in each topology snapshot period. For real-time business, the algorithm selects the shortest delay path, and for the rest of the business selects the remaining $k - 1$ number of paths in accordance with their QoS requirements. In the period close to the switching point of the snapshot, the algorithm will find $k + \Delta$ number of paths that meet the constraints for each pair of OD nodes, and each pair of OD node will choose k number of paths with least switching as a transmission path to reduce the delay jitter. The algorithm also takes into account the limited storage capacity of a satellite node, and adopts an appropriate strategy to make the satellite node unnecessary to store a complete routing table at all times. The shortcoming of the routing mechanism proposed in [39] is that a satellite node is required to be able to connect at least one terrestrial gateway when it needs to refresh the routing table.

After the satellite over satellite (SoS) architecture was proposed by Lee Jae Wook in [40], the multilayer satellite networks gradually become a hot research. The LEO satellite layer is set as layer one, and the MEO satellite layer is set as layer two. Each MEO satellite can cover several LEO satellites, and the MEO satellite is the parent node of the LEO satellite it covers. The routing information is exchanged between LEO satellites through inner orbit ISLs and interorbit ISLs, and through IOL between LEO and MEO satellites. In the architecture, each satellite has the topology of the whole network through the exchange of routing information. Hence an appropriate path can be selected according to the users' requirements. For the short distance dependent (SDD) traffic, a path without IOL has a higher priority to be chosen, while for the long distance dependent (SDD) traffic, a path with IOL will be considered first. Certain traffic balancing can be made during the routing process while satisfying the QoS requirements of the users.

Based on Lee Jae-Wook's research in [40], Chen Chao et al. proposed the group routing strategy for an LEO/MEO double-layered satellite network [41], which extended the virtual topology strategy to a double-layered satellite network for the first time. The topology fixing strategy for a satellite network in [41] utilizes the

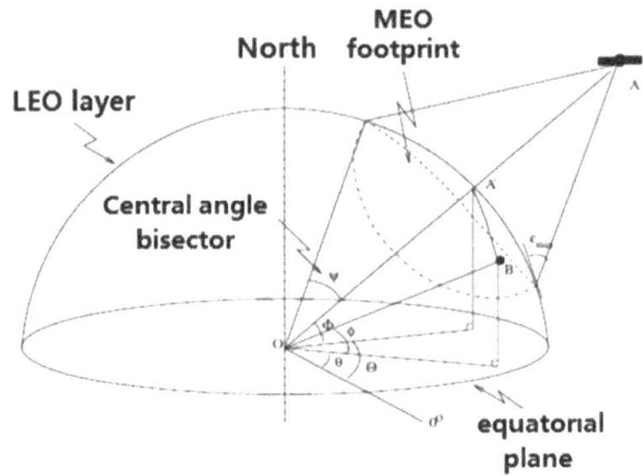

Fig. 1.4 The principle of virtual topology grouping

thought of satellite grouping of [40]. As shown in Fig. 1.4, the MEO satellite projects in an LEO layer with the minimum elevation angle ε_{min}, and the projected range is called footprint. All the LEO satellites located in the footprint form a group, and are called the group members of the MEO satellite. The MEO satellite is called the group leader satellite. With the movement of LEO and MEO satellites, the LEO satellites will leave the footprint of an MEO and enter the footprint of another MEO. The LEO/MEO group membership changes at that time. The time with the same LEO/MEO group membership is defined as a snapshot in [41]. Since the affiliation of LEO and MEO is the same, the network topology can be deemed as fixed. Thus, the continuous changing of the topology of an LEO/MEO double-layered satellite network is solidified as a series of discrete snapshots. In each snapshot, the SPF algorithm is used to find the shortest path for each OD pair. The path is calculated in advance at the terrestrial gateway and uploaded to the satellite node since the topology snapshot can be calculated also in advance. Although the topology fixing problem of multilayered satellite network is solved by the virtual topology grouping strategy in [41], the large number of snapshots in a system cycle makes snapshot switching frequent, and consequently, resulting in a huge protocol overhead. The multilayered satellite routing (MLSR) algorithm proposed in [41] is similar to the algorithm proposed in [40]. MLSR applies to LEO/MEO/GEO triple-layered satellite networks. It also divides the group according to the footprint of a higher layer satellite on a lower layer. The topology is studied hierarchically. When the group membership changes, the routing table of lower layer satellite is calculated by its group leader satellite. Since the application model of MLSR includes a global coverage MEO constellation, the system complexity is high and the robustness is low.

The virtual topology routing strategy can be extended to multilayered satellite networks. Its routing computing may be undertaken by the terrestrial gateway offline, or undertaken by a high layer satellite in real time. This greatly reduces the burden of LEO satellites and the implementation is more flexible. Compared with the virtual node routing strategy, the implementation of the virtual topology strategy is more complex and the overhead is much larger. So it is not highly adaptive to traffic change, link congestion or some technical faults. In addition, a large amount of storage space is needed to store the routing information due to the large number of snapshots divided by the virtual topology strategy in the system cycle. As a result, reducing the storage overhead has always been a big issue of the virtual topology strategy to be addressed.

1.4 Contents and Structure of the Book

1.4.1 Contents of the Book

Satellite network routing is an important part of satellite network technology. It determines the performance of a satellite network. Satellite network routing is different from terrestrial network routing. The characteristics of high-speed changing topology and constrained onboard resources make the robust routing algorithm more desirable for satellite networks. The trend of the development of satellite networking makes it necessary for it to provide QoS services to users.

This book researches satellite network robust QoS-aware routing from the aspects of constellation design, routing strategy optimization, traffic engineering, and routing algorithm design. After careful studies of the relevant literature, the author found the following deficiencies in the existing satellite network routing technology:

(1) The satellite network routing strategy is relatively simple. The application range of the virtual node routing strategy is limited. It depends a lot on constellation topology, hence is hard to be extended to multilayered satellite networks. Moreover, the robustness of virtual node topology is poor. On the other hand, the virtual topology strategy creates a large number of snapshots in a system cycle, which causes frequent link switches, large storage of routing information, rapid change of end-to-end delay, large protocol overhead, and other issues.
(2) There is little research on satellite network traffic engineering. With the trend of the integration of satellite networking and terrestrial networking, a satellite network acts as a bypass network of a terrestrial network. Thus, the loaded traffic should be considered carefully. Due to its limited onboard resources, a satellite network needs traffic engineering more than a terrestrial network in order to avoid network congestion and system crashes.
(3) The existing satellite network routing algorithms, almost all, belong to the single-objective optimization algorithm. They optimize delay or link utilization alone and ignore other QoS parameters. With the increase of satellite network applications and the growing demand for better service, choosing a most desirable path according to the users' QoS requirements is a problem to be solved using satellite network routing algorithms. When the users have more than one QoS requirement, the existing satellite network routing algorithms cannot provide such service.

In order to tackle the above-mentioned shortcomings of satellite network routing technology, this book researches the satellite constellation design, routing strategy optimization, satellite network traffic engineering, and multi-QoS-objective optimization, proposes a complete satellite network QoS robust routing technology which enables the satellite network system to provide more reliable and more stable service to the users. The routing technology proposed in this book is extensive, and makes the satellite network system achieve optimum from the point of view of both the network and the users (that is to say, it can make the network MLU minimum, while satisfy the users' multiple QoS requirements). Different from the bulk of existing research in this field, this book focuses on robust QoS routing technology for a multilayered satellite network.

1.4.2 The Main Points of the Book

(1) Constellation design and optimization. Satellite network routing is dependent on the structure of the satellite constellation; hence constellation design is related to the performance of the satellite network. A satellite constellation should be designed according to certain design objectives. Simulations are given to reveal the relationship between a satellite constellation and network transmission performance.

(2) Satellite network routing strategy. The virtual topology routing strategy is improved to reduce the snapshots in a system cycle. A snapshot merging method is proposed to eliminate snapshots that are too short. A new routing strategy integrated with the advantage of virtual topology and the virtual node strategy is proposed to enhance the scalability of the satellite network routing strategy.

(3) Satellite network traffic engineering. On the basis of the proposed satellite network routing strategy, and according to the terrestrial network traffic and density information of users around the world, satellite network traffic is predicted through time series analysis. MLU is then selected as the optimization objective. The predicted normal network traffic model is used as the optimization input, and the network link utility closure is added on the optimization. This makes the routing strategy near-optimal under the normal network traffic model. Besides, the routing strategy will not exceed the limit of overhead closure when the unpredicted traffic spike comes. Network congestion is hence reduced and the system optimum is achieved from the point of view of the network.

(4) Research of satellite network routing algorithms. Heuristic algorithms, such as the ant colony, bee colony, taboo search, and genetic algorithms, are used to solve the problem of multi-QoS objective routing optimization of a satellite network. A portfolio of different QoS objectives is optimized according to the users' requirements. Multiobjective algorithms, such as the proper equal

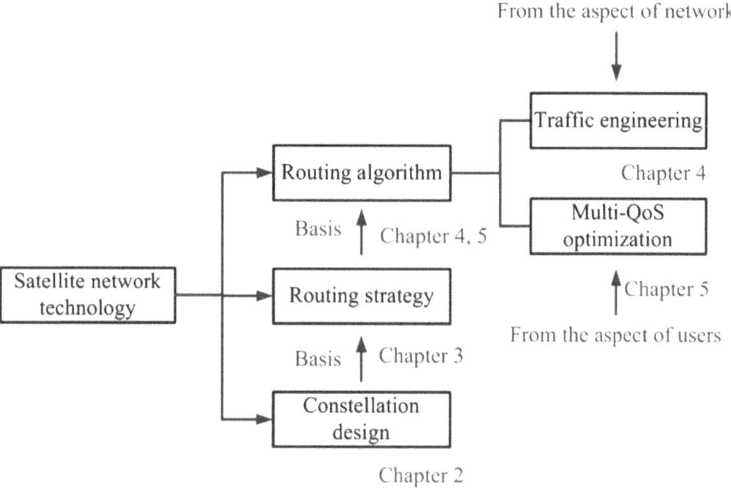

Fig. 1.5 Framework of this book

constraint (PEC) and optimum sequential methods, are also used to optimize the portfolio of different QoS objectives. The performance is compared among the heuristic routing algorithm, the multiobjective optimization routing algorithm, and the SPF routing algorithm through simulation.

1.4.3 Contents of Each Chapter

The framework of this book is shown in Fig. 1.5. Chapter 2 gives the constellation model used in this book for experiments and analyzes the influence of satellite constellation structure on the network performance. Chapter 3 gives an extensive new routing strategy based on the constellation model given in Chap. 2. The new routing strategy draws on the advantages of the virtual node and virtual topology strategy and uses a snapshot merging method. Chapter 4 first gives the satellite network traffic prediction using time series analysis according to the existing terrestrial network traffic and the density of world users, and then, optimizes for the possible peak flow of the satellite network according to the achieved traffic prediction model. Chapter 5 uses various heuristic algorithms, the PEC, and prior order methods to optimize the users' multi-QoS requirements. The performance of the multi-QoS optimization algorithm and the SPF algorithm are compared through simulations. Finally, Chap. 6 summarizes the whole book, and gives the outlook of further research on satellite networking.

References

1. Akyildiz IF, Jeong S (1997) Satellite ATM networks: a survey. IEEE Commun Mag 35(7):30–44
2. Chen C (2005) Advanced routing protocol for satellite and space networks Ph. D. thesis, Georgia Institute of Technology, Atlanta, U.S.A
3. Travis E (2001) The interplanetary internet: architecture and key technical concepts.In: Internet Global Summit, INET 2001
4. Bhasin K, Hayden JL (2002) Space internet architecture and technologies for NASA enterprises. Int J Satell Commun 20:311–332
5. Jet Propulsion Laboratory, California Institute of Technology. The NASA Deep Space Network (DSN) [EB/OL]. [2005-01-05]. http://deepspace.jpl.nasa.gov/dsn/
6. Moy J (1998) Open Shortest Path First-OSPF version 2. IETF RFC2328, Internet standard STD0054
7. Malkin G (1998) Routing Information Protocol-RIP version 2. IETF RFC2453, Internet standard STD0056
8. Broch J, Maltz DA, Johnson DB et al (1998) A performance comparison of multi-hop wireless ad hoc network routing protocols. In: Proceedings of the 4th international conference on mobile computing and networking, pp 85–97
9. Li L (2003) Space-based integrated information network. 863 aerospace technology. 8:1–14 (In Chinese)
10. Wood L (2001) Internetworking with satellite constellation. Ph. D thesis. University of Surrey, Guildford, UK
11. Elbert BR (2004) The satellite communication application handbook. Artech House Inc, Norwood
12. Chedia L, Smith K, Titzer G (1999) Satellite PCN-The ICO system. Int J Satell Commun 17:273–289
13. Leopold RJ, Miller A (1993) The Iridium communication system. IEEE Potentials 12:6–9
14. Sturza MA (1995) Architecture of Teledesic satellite system. In: Proceedings of the 4th international mobile satellite conference (IMSC'95), Ottawa, CA, pp 212–218
15. Maral G, Bousquet M (1998) Satellite communication systems. Wiley & Sons Inc, Malden
16. Fitzpatrick E (1995) Spaceway system summary. Space Commun 13(1):7–23
17. Zhou YH, Sun FC, Zhang B (2006) A time-division QoS routing protocol for Three-layered satellite networks. Chin J Comput 29(10):1813–1822
18. Rekhter Y, Li T (1995) A Border Gateway Protocol (BGP-4). IETF RFC1771, Internet standard STD0052
19. Ekici E, Akyildiz IF, Bender MD (2001) Network layer integration of terrestrial and satellite IP networks over BGP-S. In: Proceedings of IEEE GLOBECOM'2001, San Antonio, TX, pp 2698–2702
20. Ekici E, Chen C (2004) BGP-S: a protocol for terrestrial and satellite network integration in network layer. ACM/Kluwer Wireless Netw J 10(9):595–605
21. Werner M, Delueehi C, Vogel H et al (1997) ATM- based routing in LEO/MEO satellite networks with inter-satellite links. IEEE J Select Areas Commun 1:69–82
22. Franck L, Maral G (2002) Routing in networks of inter-satellite links. IEEE Trans Aerosp Electron Syst 7:902–917
23. Long F, Sun FC, Wu FG (2008) A QoS routing based on heuristic algorithm for double-layered satellite networks. In: IEEE World Congress on computational intelligence (WCCI'2008), vol 6, pp 1866–2008
24. Ekici E, Akyildiz IF, Micheal DB (2000) Datagram routing algorithm for LEO satellite networks. In: Proceedings of IEEE INFOCOM'2000, Tel-Aviv, Israel, pp 500–508
25. Werner M (1997) A dynamic routing concept for ATM-based satellite personal communication networks. IEEE J Sel Areas Commun 8:1636–1648

26. Hashimoto Y, Sarikaya B (1998) Design of IP-based routing in a LEO satellite network. Proceedings of the 3rd international workshop on satellite-based information services, Mobicom'1998, vol 10, pp 81–88
27. Henderson TR, Katz RH (2000) On distributed geographic based packet routing for LEO satellite networks. In: Proceedings of IEEE GLOBECOM'2000, San Francisco, CA, pp 1119–1121
28. Ekici E, Akyildiz IF, Bender MD (2001) A distributed routing algorithm for datagram traffic in LEO satellite networks. IEEE/ACM Trans Networking 2:137–147
29. Uzunalioglu H (1998) Probabilistic routing protocol for Low Earth Orbit satellite networks. In: Proceedings of IEEE International conference on communications (ICC'1998), Atlanta, GE, pp 88–93
30. Chan TH, Yeo BS, Turner LF (2003) A localized routing scheme for LEO satellite networks. In: Proceedings of AIAA 21st international communication satellite systems conference and exhibit (ICSSC'2003), Yokohama, Japan, pp 2357–2364
31. Lee WT, Kung LY (1995) Binary addressing and routing schemes in the Manhattan Street Network. IEEE/ACM Trans Networking 1:26–30
32. Chen C (2003) A QoS based routing algorithm in multimedia satellite networks. In: Proceedings of IEEE 58th vehicular technology conference, Orlando, FL, pp 2703–2707
33. Nguyen HN, Jukan A (2000) An approach to QoS-based routing for Low Earth Orbit satellite networks. In: IEEE Global Telecommunications Conference (GLOBECOM'2000), vol 2. San Francisco, CA, pp 1114–1118
34. Sanctis MD, Cianca E, Ruggieri M (2003) IP-based routing algorithms for LEO satellite networks in near-polar orbits. In: Proceedings of the IEEE 2003 aerospace conference, Big sky, MT, vol 3, pp 1273–1280
35. Hu Y, Li VK (2001) Logical topology based routing in LEO constellations. In: Proceedings of IEEE international conference on communications (ICC'2001), vol 5. St. Petersburg, Russia, pp 3172–3176
36. Kim YS, Park WJ (1998) Traffic load balancing in Low Earth Orbit satellite networks. In: Proceedings of IEEE International conference on computer communications and networks, vol 10. Lafayette, LA, pp 191–195
37. Chang HS, Kim BW, Lee CG et al (1995) Topology design and routing for Low Earth Orbit satellite networks. In: Proceedings of IEEE global telecommunications conference (Globecom'1995), vol 11. Singapore, pp 529–535
38. Chang HS, Kim BW, Lee CG et al (1998) FSA based link assignment and routing in Low Earth Orbit satellite networks. IEEE Trans Veh Technol 8:1037–1048
39. Vidyashankar VG, Ravi P, Hosame A (1999) Routing in LEO based satellite networks. In: Proceedings of IEEE Emerging technologies symposium on wireless communications and systems, vol 4. Richardson, TX, pp 12–13
40. Lee J, Kang S (2000) Satellite over satellite (SOS) network: a novel architecture for satellite network. In: Proceedings of IEEE INFOCOM'2000, vol 3, Issue 1. Tel-Aviv Israel, pp 315–321
41. Chen C, Ekici E, Akyildiz IF (2002) Satellite grouping and routing protocol for LEO/MEO satellite IP networks. In: Proceedings of the 5th ACM International workshop on wireless mobile multimedia (WoWMoM'2002), vol 9. Rome, Italy, pp 109–116

Chapter 2
Satellite Network Constellation Design

2.1 Introduction

Satellite constellation design plays an important role in satellite network robust QoS routing technology. Although robust QoS routing technology should not depend on constellation design, constellation design can affect the system cost, the effective communications of the entire network, and the effective and convenient management of the satellite network. At the same time, it is the basis of the implementation of routing strategies and routing algorithms. Constellation design is defined as follows in [1]: distribution of satellites of similar types or with similar functions in similar or complementary orbits to accomplish a specific task under shared control. The thought of using satellite constellation to provide wireless communications service can be traced back to 1945. British scientist Clarke published a paper in Wireless World, and first proposed a scheme of using three GEO satellites to provide continuous coverage for the equatorial region [2]. This scheme can be deemed as the first constellation design scheme.

Although a GEO satellite can cover low latitude regions well, it cannot provide good service for the high latitude areas. Besides, the power requirement of the terrestrial terminal and the quality requirement of the receivers are very high in a GEO communication system, which means the terrestrial terminal is hard to be made handheld. This is because the orbit of GEO satellite is extremely high which leads to a long round trip time (RTT) and a heavy link loss. For some business with high real-time requirement, GEO satellites sometimes are difficult to guarantee the quality of service. The emergence of non-geostationary orbit satellites such as LEO, MEO satellites greatly compensates the disadvantage of GEO satellites. Compared with GEO satellites, LEO and MEO satellites have smaller communication delays and link loss. The application range of LEO and MEO satellites is larger than GEO satellites. Regional coverage, intermittent coverage, and global seamless coverage can be achieved by adjusting the parameters of LEO and MEO constellations. LEO and MEO satellites make constellation design more complex, and non-geostationary orbit constellation design has gradually become the focus of constellation design.

F. Long, *Satellite Network Robust QoS-aware Routing*,
DOI: 10.1007/978-3-642-54353-1_2,

Since the 1960s, great achievements have been made in terms of non-geostationary orbit constellation design. Some representative examples are as follows: LEO polar orbit constellations, the δ Constellation proposed by Walker, the Rosette Constellation proposed by Ballard, the Ω Constellation and the σ Constellation, etc. The researches of Walker, Ballard, Rider, Hanson et al. are more influential. Their proposed constellation design schemes [3–11] have been widely adopted by later researchers, and become classic in non-geostationary orbit constellation design [12].

Although the efforts of Walker et al. have solved the inherent disadvantages of GEO satellite constellations, such as limited coverage, long communication delays, and large link loss, the drawbacks of vulnerability and limited constellation resources are still unsolved. The emergence of satellite over satellite (SOS) architecture [13] makes the collaboration between different layers come true. It enhances the invulnerability of satellite network systems, and lightens the burden of routing of LEO satellites, and also makes the design of non-geostationary orbit satellite constellation more complicated. Constellation design has thus been shifted from a single layer to multilayers. The representative works on multilayered constellation design are also presented in [14–17]. These works not only take into account constellation coverage, but also consider constellation failure robustness, the total system cost, and many other factors.

The principles and methods of non-geostationary orbit constellation design are systematically expounded in [12], and two classic constellation design methods, the Orthogonal Circular Orbit Constellation Design and the Common Ground Track Regional Constellation Design, are proposed. The Orthogonal Circular Orbit Design uses geostationary satellites and polar-orbiting satellites to provide service for terrestrial users. This design takes full advantage of continuous band coverage of geostationary satellites in low latitudes and continuous crown coverage of polar orbit satellites. It considers the population-weighted average elevation angle, latitude average elevation, and also the distribution of average elevation angle. The design improves the coverage in mid- and low latitude regions, especially the low latitude regions, while keeping good coverage in high latitude areas. This also makes the coverage characteristics of the whole constellation more consistent with the global population latitude distribution characteristics, and improves the overall coverage performance.

The disadvantage of the Orthogonal Circular Orbit Design is that, when the total number of constellation satellites, the number of orbital planes and the number of satellites in each orbit are all fixed, the orbit altitude required by the Orthogonal Circular Orbit Design to achieve global coverage is higher than polar orbit constellations. The Common Ground Track Regional Constellation Design draws on the idea of the common ground track continuous coverage design proposed in [18]. The constellation design in [19] is of this kind. It considers the regional population distribution of China and can provide seamless coverage for the whole territory of China.

A simple encoding identifier that can completely describe the parameters of common ground track constellations was first proposed in [18]. The equivalent condition between common ground constellations is then discussed, and the common optimization design method to realize regional satellite mobile communication is given. In the end, it makes an adjustment of the parameters of a constellation program based on the time varying characteristics of business distribution density. The parameters are further optimized by the genetic algorithm.

Besides, there is a lot of other literature on multilayered constellation design. These constellation designs are conceived from the point of view of robustness, system cost, and coverage performance, etc. This book does not focus on constellation design. The constellation design mentioned in this chapter is intended to provide an application model for the implementation of satellite network routing schemes and robust QoS routing algorithms, and a useful research method for multilayered constellation design. Therefore, the constellation design objectives considered in this chapter are as follows [19]:

(1) 100 % earth coverage by LEO layer satellites.
(2) 100 % earth coverage by MEO layer satellites.
(3) 100 % coverage of LEO satellites by MEO layer satellites.
(4) 100 % coverage of mid-low regions of the earth by GEO layer satellites.
(5) The access satellite link duration time of any user (no switching) $\geq$5 min.
(6) The duration of continuously covering one ground station by one satellite $\geq$8 min.
(7) The system cycle of a constellation is minimum.
(8) The number of satellites and orbits is minimum.

2.2 Principle of Constellation Design

2.2.1 Constellation Structure Selection

The satellite network system mentioned in Sect. 1.2 can be divided into the geostationary orbit system and the non-geostationary orbit system. The altitude of a geostationary orbit is 35,800 km, so the value of RTT is large, and the transmitting power required by a ground personal terminal is relatively high. Currently, most of the satellite systems in use are in non-geostationary orbits. The non-geostationary orbit satellites can be divided into circular orbit satellites and elliptical orbit satellites according to the orbit type. An elliptical orbit satellite has good coverage of high latitude areas. However, the elliptical orbital inclination must be 63.14 in order to make the satellite apogee fixed under disturbance conditions, which is very unfavorable to the coverage of mid-low latitude areas. As a result, only circular orbit satellite constellations are considered in this book. The inclination of a circular orbit can be chosen between 0 and 90°. A circular satellite network system can be either an LEO or MEO satellite network according to the orbital altitude.

In the composition of a multilayered constellation, usually the number of GEO satellites is the smallest. The orbital altitude of GEO satellites is the highest and the link loss is the heaviest, hence the required user terminal's effective isotropic-radiated power (EIRP) and G/T[1] value is the highest. The number of MEO and LEO satellite is usually larger. The orbital altitudes of MEO and LEO satellites are lower and the link loss is relatively small, which makes lower EIRP and G/T requirements of the ground station. Compared with GEO satellites, LEO and MEO satellites are more conductive to the communication between ground users and the satellite. However, in order to achieve the design objectives of (1)–(3), the required number of LEO and MEO satellites are much larger, which makes a higher cost than a GEO system. At the same time, LEO and MEO satellites also have the disadvantages of more complex satellite switching control and the serious Doppler Effect. These problems are particularly prominent in an LEO constellation. The combination of GEO MEO and LEO satellites can effectively make up for the disadvantages of a single-layered satellite constellation.

In order to test the performance of the various constellation structures, we designed two simulation experiments. Both of them adopt the satellite network routing strategy and the routing algorithm proposed in [14]. The first experiment compared the end to end delays of satellite nodes of an LEO constellation, an LEO/MEO constellation, and an LEO/MEO/GEO constellation all under a low network load. The LEO single-layered constellation adopts the constellation parameters of Iridium [20], the LEO/MEO double-layered constellation uses the parameters of Iridium and ICO [21], and the LEO/MEO/GEO triple-layered constellation adopts the parameters of Iridium, ICO and three evenly distributed GEO satellites. The second experiment compared the end to end delays of three constellations all under a high network load. In our simulation, link utilization less than 95 % is deemed as low network load, and conversely, high network load. Link utilization is set by adjusting the link background traffic. Beijing (116°23′E, 39°54′N) is chosen as the central ground station in this simulation, the latitudes of the other 12 test nodes remain the same, and their longitude interval is 15° eastward from Beijing successively. The end to end delay between the nodes is the average value of the 24-h simulation time, and Figs. 2.1 and 2.2 show the simulation results.

As shown in the figures, the performance of the LEO single-layered constellation is superior to that of both the LEO/MEO double-layered satellite constellation and the LEO/MEO/GEO triple-layered constellation for the transmission delay plays a decisive role under the low network load (link utilization 90 %).

[1] The G/T value of ground station performance index is an important technical indicator which reflects the performance of a ground station receiving system. Where G is the receiver antenna gain, T is the equivalent noise temperature of receiving system noise performance. A larger G/T value reflects a better performance of the ground receiving system. At present, ground stations with G/T $\geq$ 35 dB/K are defined as A-type standard stations, ground stations with G/T $\geq$ 31.7 dB/K are defined as B-type standard stations, while the stations with G/T $<$ 31.7 dB/K are called nonstandard stations internationally.

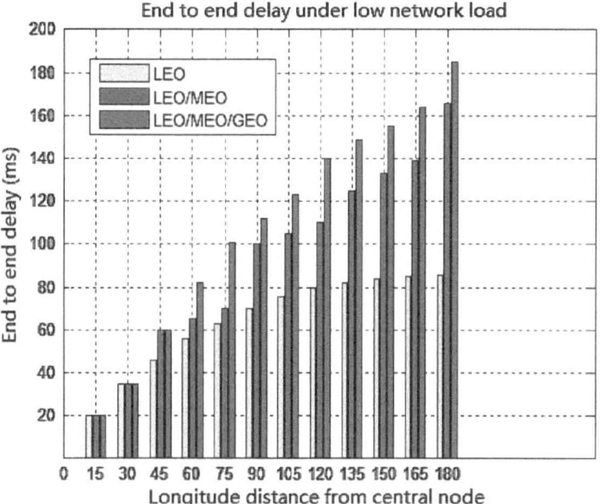

Fig. 2.1 End to end delay when link utilization at 90 %

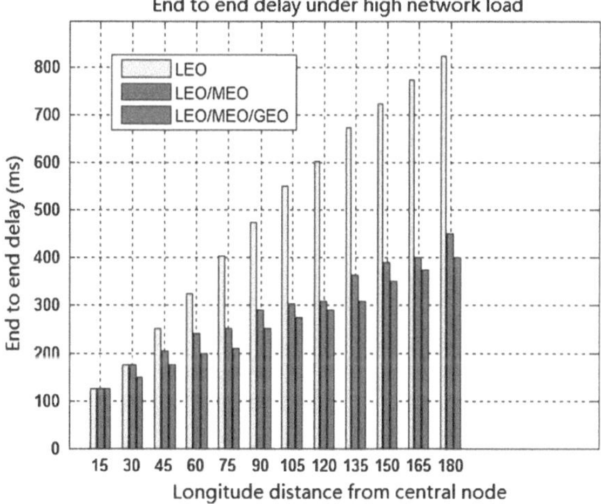

Fig. 2.2 End to end delay when link utilization at 98 %

However, when the network load is high (link utilization 98 %), the queuing delay, processing delay, and the probable congestion delay play a decisive role. At this time, the advantages of high bandwidth, multiple path, high processing ability, and high reliability of double-layered and triple-layered constellations start to show, and make their performances superior to the single-layered constellation.

As known from the requirement analyses and service experience of the operational constellations, the business types of a satellite network are varied and the traffic is huge [20, 21]. So an LEO/MEO/GEO triple-layered satellite constellation is chosen as our experimental model. Since there does not exist a constellation model that fits all business types, the performance of the LEO/MEO/GEO triple-layered satellite constellation may be inferior to that of a double or single-layered

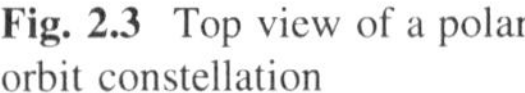

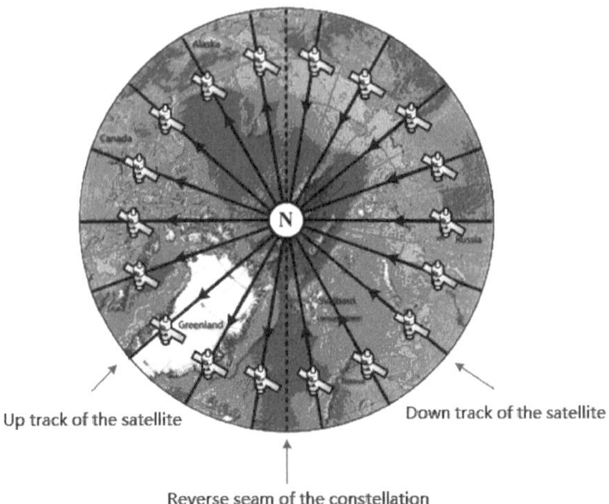

Fig. 2.3 Top view of a polar orbit constellation

satellite constellation under certain circumstances. The constellation design must be in compliance with user requirements and the business type.

2.2.2 Orbit Type Selection

This book discusses circular orbit constellations only. As mentioned before, circular orbits can be divided into polar orbits and inclined orbits. The orbital inclination angle of a polar orbit is equal or close to 90°. A polar orbit is so-called because the satellite in it periodically crosses the north and south poles. The orbital inclination angle of an inclined orbit is less than 90°. Among the operational constellation systems, both Iridium [20] and Teledisic [22] use a polar orbit, while the Globalstar [23] constellation and the ICO [21] constellation adopt an inclined orbit.

A top view and a lateral view of a polar orbit constellation are shown in Figs. 2.3 and 2.4, respectively. Figure 2.3 shows the polar orbit constellation seen from above the Arctic polar region. Satellites east of the North Pole run up approaching the North Pole and the satellites west of the North Pole run along their tracks away from the North Pole. The satellites in the track 10° northeast of the North Pole and the satellites in the track 10° northwest of the North Pole run in opposite directions. Similarly, the satellites in the track 10° southeast of the North Pole and the satellites in the track 10° southwest of the North Pole also run oppositely. The dotted line in Fig. 2.3 shows the reverse seam of the polar orbit constellation track, satellites on both sides of the seam running reversely. Figure 2.4 shows a side view of a polar-orbiting constellation. Every single polar-orbiting satellite maintains four ISLs, two of which connect the adjacent satellites within the same orbit and the other two connect the adjacent satellites in two adjacent tracks. ISLs connecting the adjacent

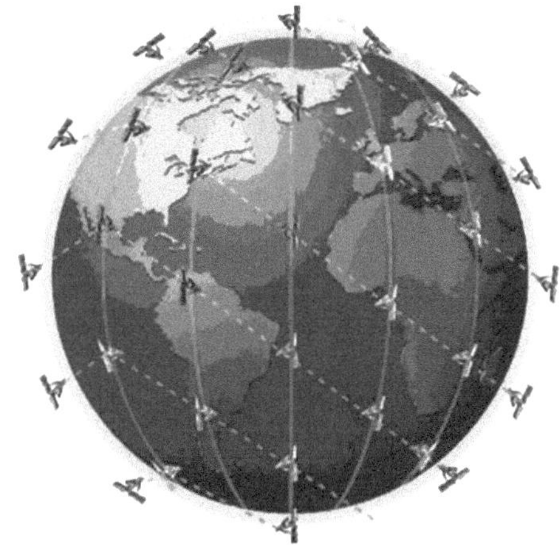

Fig. 2.4 Side view of a polar orbit constellation

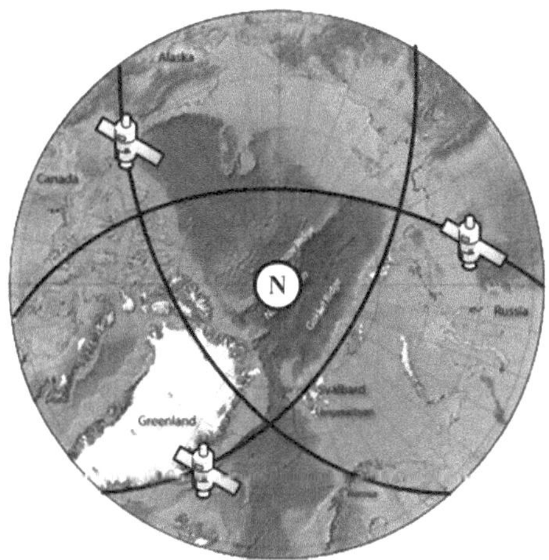

Fig. 2.5 Top view of an inclined orbit constellation

satellites in the same orbit are permanent links. ISLs connecting the adjacent satellites in adjacent orbits will be temporarily closed when crossing the Polar Regions, for the antennas cannot track the neighboring satellites in time during the high-speed swap of satellite relative positions. Satellites in both sides of the reverse seam maintain only three ISLs. This is because the satellites in both sides of the seam move reversely at an extremely high speed, and not suitable for the establishment of ISL. Although a polar orbit constellation has the disadvantage of sparse coverage of the low latitude regions, it is very suitable for the implementation of the virtual node routing strategy, which will be discussed in the following chapters.

Top and side views of an inclined orbit constellation are shown in Figs. 2.5 and 2.6, respectively. Figure 2.5 shows a polar orbit constellation seen from above the Arctic polar region. The constellation shown in Fig. 2.5 was proposed by Walker first, and named the δ Constellation for three satellite tracks compose the Greek

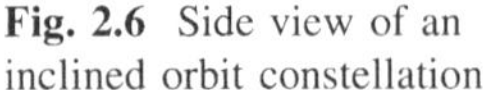

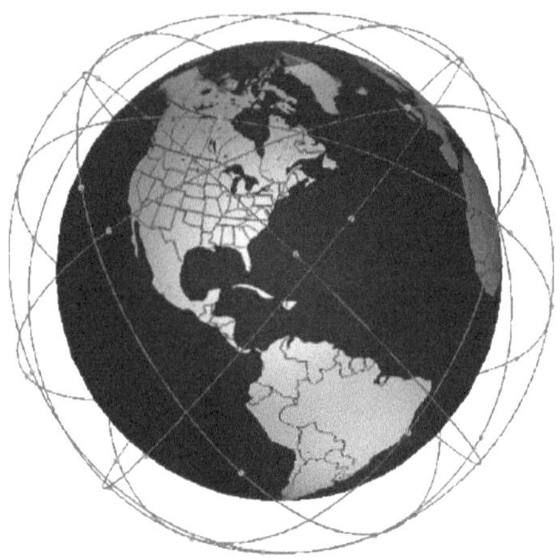

Fig. 2.6 Side view of an inclined orbit constellation

alphabet Δ as shown in the top view. The side view of the inclined orbit constellation is shown in Fig. 2.6. In the inclined orbit constellation, each satellite keeps two ISLs connecting two adjacent satellites on the same orbital plane. These two ISLs are permanent links as in a polar orbit constellation. Since the distance between two adjacent satellites in two adjacent orbital planes changes greatly and rapidly in an inclined orbit constellation, ISLs are usually established near the intersection point of two orbital planes and information is then exchanged when the link is established. Although uniform global coverage can be achieved by an inclined orbit constellation through proper arrangement, its coverage area is not regular and the changes of connections are more complex than a polar orbit constellation. Therefore, the virtual node routing strategy (which will be discussed in the following chapters) is not suitable for an inclined orbit constellation.

In practice, different layered satellites usually collaborate with one another. The LEO-layered satellites are usually used as entrance/exit satellites, which connect and communicate with terrestrial gateways directly. A polar orbit constellation is usually chosen for the LEO layer for the easy implementation of the virtual node strategy. MEO-layered satellites are usually used as the manager of the LEO satellites. The inclined orbit constellation is sometimes used in the MEO layer to provide a uniform coverage of the Earth and also the LEO-layered satellites.

2.2.3 Selection of Orbit Altitude

The orbit altitude is an important factor in the design of a satellite constellation. All of the design goals mentioned in Sect. 2.1 are directly related to the orbit altitude. In addition, EIRP and G/T value of ground terminals are also related to the orbit altitude. A higher orbit altitude means higher requirements of EIRP and G/T value. However, the required number of satellites to achieve the goals (1)–(8)

Table 2.1 Possible choice of LEO orbit altitude

N	K_1	h (km)
1	13, 14	1248, 896
2	25, 27, 29	1461, 1070, 725

is fewer with a higher orbit altitude, which means a lower construction cost. The choice of orbit altitude usually requires a compromise between the construction cost and the quality of service; meanwhile, the following two environmental restrictions are also needed to be considered.

(1) The influence of the Earth's atmosphere: In general, the orbit altitude of an LEO satellite should not be lower than the altitude of the top of the atmosphere, or the oxygen atoms of the upper atmosphere will erode the body of the satellite seriously, and shorten the service life of the satellite. At the same time, the disturbance of the atmosphere will also affect the normal operation of the satellite. The disturbance and damping of the atmosphere can be ignored only when the orbit altitude is greater than 1000 km.
(2) The influence of the Van Allen belts: As mentioned in Sect. 1.2, the Van Allen belts are named after their discover Van Allen. They are two radiation belts surrounding the Earth. They are composed of high-energy charged particles, and present strong electromagnetic radiation. The Van Allen belts are divided into the inner belt and the outer belt located at an altitude of 1,500–5,000 km and 13,000–20,000 km, respectively. When designing the LEO and MEO constellations, the altitude of an orbit should avoid being located in the two Van Allen belts so as to keep away from the strong electromagnetic radiation of these two belts.

For the ease of computing, the constellation should be able to return to the initial state after a period of time. If the constellation is made up of LEO-layered satellites and MEO-layered satellites, the running cycle of LEO satellites T_L and MEO satellites T_M should satisfy the following equation:

$$T_L \times K_1 = T_M \times K_2 = T_E \times N,$$

where T_E is the rotation period of the Earth, and K_1, K_2, N are all integers.

According to the Kepler's third law:

$$T = 2\pi(R_E + h)\sqrt{(R_E + h)/GM}, \tag{2.1}$$

where T is the running cycle of satellite, R_E is the radius of Earth, h is the satellite orbit altitude, G is the universal gravitational constant, M is the weight of Earth. $R_E = 6378.14$ km, $G = 6.67 \times 10^{-11}$ m^3 kg^{-1} s^{-2}, $M = 5.97 \times 10^{24}$ kg.

For the LEO satellites, according to the environmental constraint on the satellite orbit and the definition of the LEO satellite orbit in Sect. 1.2, the possible choice of LEO orbit altitude can be calculated by Eq. (2.1) as shown in (Table 2.1).

Table 2.2 Possible choice of MEO orbit altitude

N	K_2	h (km)
1	4, 5	10390, 8042
2	7	11912

$N = 1$ means the constellation will return to the initial state after 24 h of running of the LEO-layered satellites, while $N = 2$ means the constellation will return to the initial state after 48 h of running of the LEO-layered satellite. For the ease of computing, the situations when $N > 2$ are not considered in this book.

Similarly, for the MEO satellites, the possible choices of orbit altitudes as shown in (Table 2.2).

Besides the environmental constraint and the repetitiveness of the constellation initial state, the running cycle of the entire constellation should also be considered when choosing the orbit altitudes of LEO and MEO satellites. If the terrestrial gateway is considered, the system cycle of the multilayered constellation should be the least common multiple of the running cycle of the satellites of each layer and the rotation period of the Earth. And if only the space-based network is considered, the system cycle of the multilayered constellation should be the least common multiple of the running cycle of the satellites of each layer. Since the running period of the GEO satellites equals the rotation period of the Earth, in order to satisfy the design goals (7) mentioned above, the least common multiple of the running period of LEO and MEO-layered satellites should be minimum and should not exceed 24 h. Therefore, 896 km is chosen as the LEO orbit altitude, and the running cycle is 12/7 h accordingly. 10390 km is chosen as the MEO orbit altitude and accordingly the running cycle is 6 h. The system cycle of the whole constellation is hence 24 h.

2.2.4 Selection of the Numbers of Orbits and Satellites

2.2.4.1 Selection of the Number of LEO Orbital Planes and Satellites

As mentioned in Sect. 2.2.2, polar orbit LEO satellites are selected as our LEO-layered satellites. These satellites are evenly distributed in different LEO orbital planes, and the LEO orbital planes are evenly distributed according to their longitudes. The LEO layer of the constellation considered in this section is like the LEO polar constellation shown in Fig. 2.4. The selection of the number of LEO orbital planes should be considered first. The global coverage is the only factor needed to be considered in the LEO layer constellation design. The coverage of the LEO polar orbit constellation is dense in high latitude areas and sparse in low latitude areas. Therefore, if the LEO constellation can completely cover the equatorial area, it can achieve global coverage. Figure 2.7 shows the cross-sectional profile of an LEO satellite's coverage of the equatorial region.

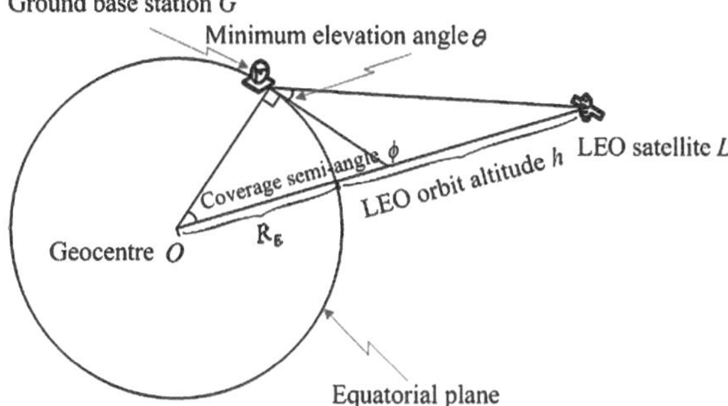

Fig. 2.7 An LEO satellite's coverage of to the equatorial region

As shown in Fig. 2.7, the longitudinal range of an LEO satellite's coverage area can be obtained by calculating the coverage semi-angle ϕ. The required number of LEO orbital planes for full coverage of the equatorial region can hence be calculated. Given that $OG = R_E = 6378.14$ km, $OL = h + R_E = 895.5$ km $+$ 6378.14 km $=$ 7273.64 km, the minimum elevation angle $\phi = 10°$. GL can first be obtained by the law of cosines:

$$OL^2 = OG^2 + GL^2 - 2OG \cdot GL \cdot \cos \angle OGL. \tag{2.2}$$

Substitute OL, OG, $\angle OGL$ into Eq. (2.2), according to the root formula, GL can be obtained by omitting a negative root:

$$GL = 2560.1 \text{ km}.$$

And then, ϕ can be obtained by the law of sines.

$$GL/\sin\phi = OL/\sin\angle OGL. \tag{2.3}$$

Substitute GL, OL, $\angle OGL$ into Eq. (2.3), ϕ can be obtained by omitting an obtuse angle solution.

$$\phi = 20.25°.$$

As a result, an LEO satellite can cover $2\phi = 40.5°$ longitudinal range of the equatorial region. That is to say, $\lceil 360^\circ/4\phi \rceil = 5$ LEO orbital planes are needed to achieve the full coverage of the equatorial region.

The selection of the number of LEO satellites in each orbital plane is similar to the selection of the number of LEO orbits. Assuming that all of the LEO satellites have the same orbit altitude, an LEO orbital plane's full coverage of the corresponding longitudinal circle is the only factor needed to be considered to achieve full coverage of the Earth. Figure 2.8 shows the longitudinal profile of an LEO satellite's coverage of the certain longitudinal circle.

As shown in Fig. 2.8, an LEO satellite's coverage range can be obtained by calculating the coverage semi-angle ϕ'. The number of LEO satellites needed to fully cover that longitudinal circle can be thus obtained. Similar to the deduction

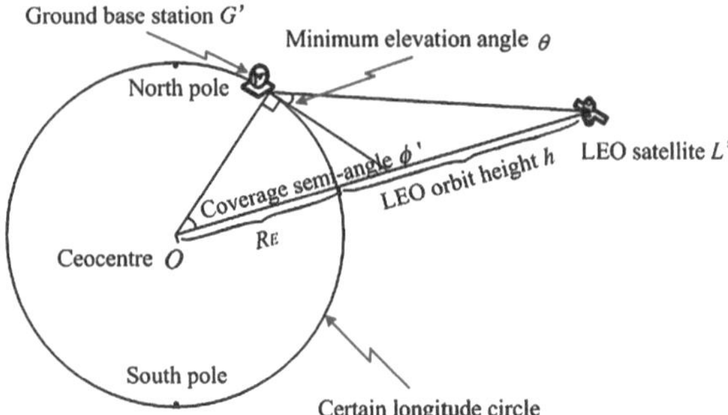

Fig. 2.8 Coverage of LEO satellite to certain longitude circle

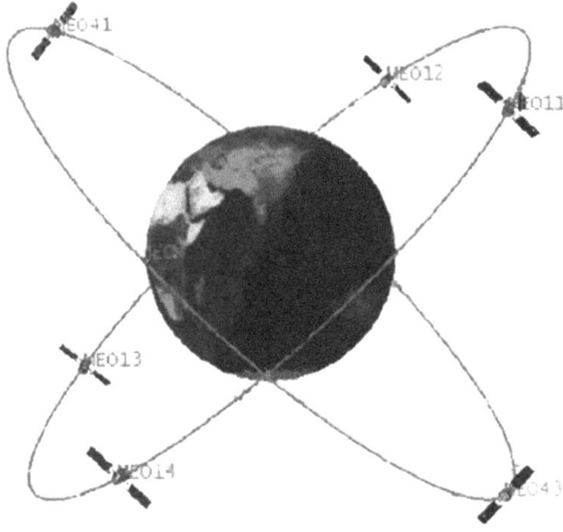

Fig. 2.9 Schematic diagram of the ICO constellation

of ϕ, we can get $\phi' = 20.25°$. That is to say, an LEO satellite's longitudinal coverage range is $2\phi' = 40.5°$. Therefore, $\lceil 360^{\circ}/2\phi' \rceil = 9$ LEO satellites are needed to fully cover the whole longitudinal circle.

2.2.4.2 Selection of the Number of MEO Orbital Planes and Satellites

For the MEO layer satellite orbit design, the two MEO orbital planes of the ICO constellation system are adopted in this book. Both of these two MEO orbital planes form a 45° included angle with the equatorial plane. Their phase difference is 180°. Since the ICO constellation is a commercial system already in use, its orbit layout is reasonable, and is adopted as our MEO orbits in the multilayered constellation. The ICO constellation is shown in Fig. 2.9. The constellation includes two orthogonal orbital planes each with five evenly distributed MEO satellites (one backup MEO satellite is not included). Each MEO satellite keeps two ISLs connecting two adjacent MEO satellites. Meanwhile, each MEO satellite establishes an ISL with the adjacent MEO satellite in the adjacent orbital plane to

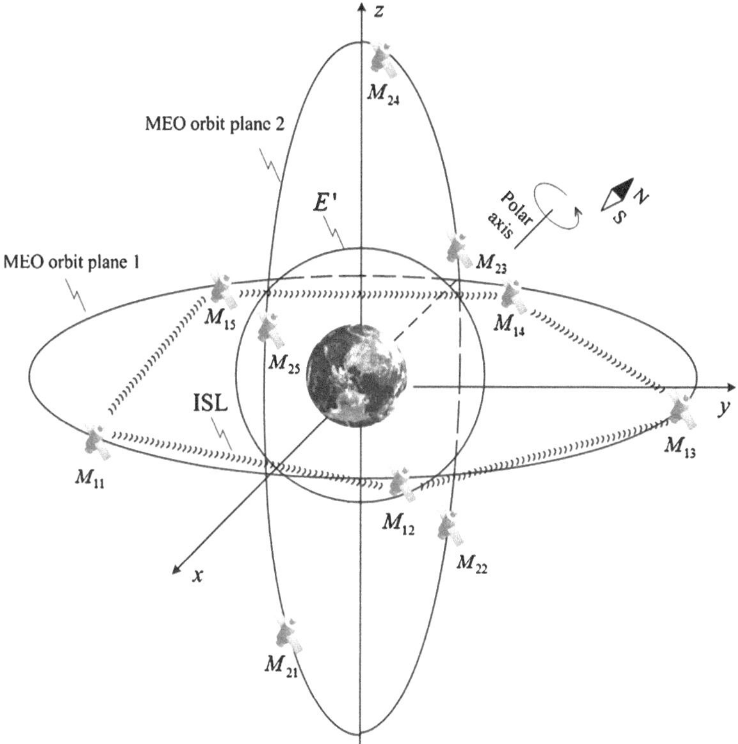

Fig. 2.10 The ICO planes in new coordinate

exchange data when it bypasses the cross point of two orbital planes. The ICO system was initially used to provide satellite mobile communications service for navigation management and distress alerting in the maritime industry.

Below we will investigate if the arrangement of orbits and satellites of the ICO system can satisfy the design goals (2) and (3). Since the polar orbit constellation design is used in the LEO layer, all of the tracks of the LEO layer form a sphere concentric with the Earth, referred to as Ball E'. Intuitively, if the MEO layer satellites can completely cover the surface of E', they can also fully cover the surface of the Earth. The process of proving this is omitted.

For the ease of computing, only one MEO orbital plane is used as the *xoy* plane to establish a coordinate system, as shown in Fig. 2.10. First, let us investigate the orbital plane 1 satellite coverage of E'. Using plane *xoy* as the cross section, we draw a sectional view of the coverage of MEO satellite M_{1x} over E' as shown in Fig. 2.11. The MEO satellite coverage of E' can be obtained by calculating the coverage semi-angle ϕ of the MEO satellite over E' As mentioned in Sect. 2.2.3, the radius of the LEO orbit $R_L = OL = 7273.64$ km, the radius of the LEO orbit $R_M = OM = 6378.14\text{ km} + 10390\text{ km} = 16768.14$ km. If the minimum

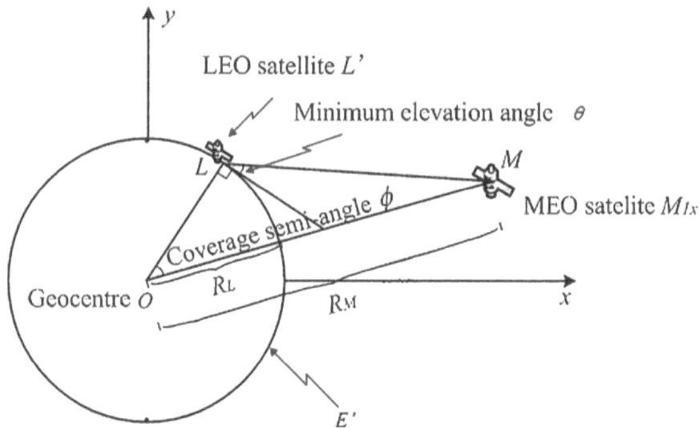

Fig. 2.11 The cross-sectional view of the MEO coverage of the LEO

elevation $\phi = 10°$, we have $\angle OLM = 90° + \phi = 100°$. *LM* can be obtained by the law of cosines as follows:

$$OM^2 = OL^2 + LM^2 - 2OL \cdot LM \cdot \cos \angle OLM. \tag{2.4}$$

Substitute *OL*, *OM* and $\angle OLM$ into Eq. (2.4), according to the root formula, LM can be got by omitting one negative root.

$$LM = 13898 \text{ km}.$$

ϕ can be got according to the law of sines.

$$LM/\sin\phi = OM/\sin\angle OLM. \tag{2.5}$$

Substitute *LM*, *OM* and $\angle OLM$ into Eq. (2.5) and omit an obtuse angle, we have:

$$\phi = 54.72°.$$

As a result, an MEO satellite can cover a $2\phi = 109.44°$ spherical range. To achieve the full coverage of E' on the *xoy* plane, at least $\lceil 360^{\circ}/2\phi \rceil = 4$ MEO satellites are needed in MEO orbital plane 1. Similarly, to achieve the full coverage of E' on the *xoz* plane, a minimum $\lceil 360^{\circ}/2\phi \rceil = 4$ MEO satellites are needed in MEO orbital plane 2. So, the arrangement of orbits and satellites of the ICO constellation system can achieve the design objectives (2) and (3).

2.2.5 Selection of Model Parameters of a Multilayered Constellation

Based on the theoretical analysis of various constellation parameters in Sects. 2.2.1–2.2.4, an LEO/MEO/GEO triple-layered constellation is designed for our further experiment and the parameters of this constellation are as follows.

Table 2.3 The parameters of the designed experimental constellation

	Orbit height (km)	Running cycle (h)	Number of satellites	Orbit inclination
LEO	895.5	12/7	6 × 11	90°
MEO	10390	6	2 × 5	45°
GEO	36000	24	1 × 3	0°

Fig. 2.12 The initial state of subsatellite points of LEO/MEO constellation

Through theoretical analysis, goals (1)–(4) and (7) can be achieved by the triple-layered constellation as shown in Table 2.3. Taking into account the survivability and robustness of the constellation system, the multilayered constellation does not stick to the requirement of goal (8), but adds a small number of redundant satellites. The goals of (5) and (6) are not involved in the theoretical analysis, and will be verified through the simulations in the following section. The constellation shown in Table 2.3 is tentatively named *Tr* in this book.

2.3 Simulation Analysis of Constellation Design

Figure 2.12 shows initial subsatellite points of the LEO and MEO satellites in constellation *Tr* through STKv5.0. The subsatellite points of GEO satellites are not shown in the graph for their relative positions are fixed. Through the simulation, we know that the LEO satellite constellation can achieve global coverage of the Earth in its whole running cycle, which tallies with our theoretical analysis. Since the LEO-layered satellites are usually used as the interface satellites of a terrestrial base station and a space-based network, to satisfy the design goal (6), we only need to investigate if the continuous coverage time of an LEO satellite over a ground base station is no less than several minutes (for an LEO satellite, the time is usually 8 min). The LEO-layered satellites of a multilayered constellation have the shortest time to cover the terrestrial base station. Therefore, the continuous

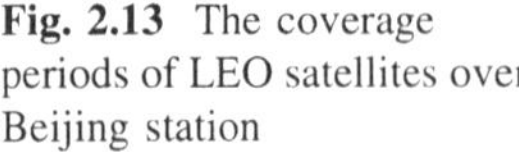

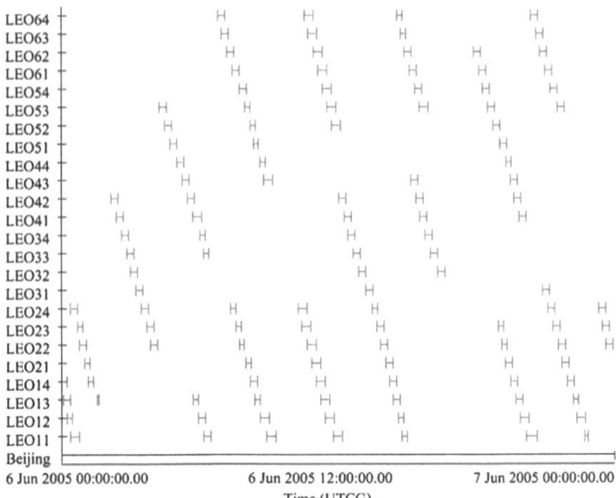

Fig. 2.13 The coverage periods of LEO satellites over Beijing station

coverage time of LEO-layered satellites is critical to the constellation design. If the continuous coverage time is too short, the frequent switching of the interface link will lead to a large protocol overhead and link instability.

Now, let's find out the time during which the LEO-layered satellites continuously cover the ground base station. Taking the ground base station in Beijing as an example, STK v5.0 is used as our simulation tool, the simulation time is set from 0:00 June 6, 2005 to 0:00 June 7, 2005. There are $6 \times 11 = 66$ LEO satellites involved in this simulation. Figure 2.13 shows the coverage time of LEO satellites numbered 1–4 in each orbital plane over the Beijing station. The simulation result shows that at least three LEO satellites can cover the Beijing station, and the maximum number is seven each day. The least continuous coverage time of LEO-layered satellites over the Beijing station is 496 s, which meets the requirement of goal (6).

Theoretically, if an LEO satellite can cover a ground base station, the transmission link between the ground station and the LEO satellite can be established. The access link duration should be equivalent to the coverage time of the LEO satellite over this ground base station. Actually, in order to ensure the link quality, the received power value of the ground station is required to be greater than a certain threshold value. That is to say, some ground stations need to switch to a new interface satellite to get higher received power even if they are still located in the coverage area of their old interface satellite. This leads to the duration time of the access link a little less than the coverage time of the current satellite. Still take Beijing station as an example, the access link duration time is checked in the simulation. Figure 2.14 shows the access link duration time of the 66 LEO satellites covering the Beijing station, where the horizontal axis represents the simulation time and the vertical axis is the minimum link duration time of all the LEO satellites covering the Beijing station at the current simulation time. This is based on the assumption that the receiving antenna of the Beijing station has a gain of 44 dB in the direction of the working satellites.

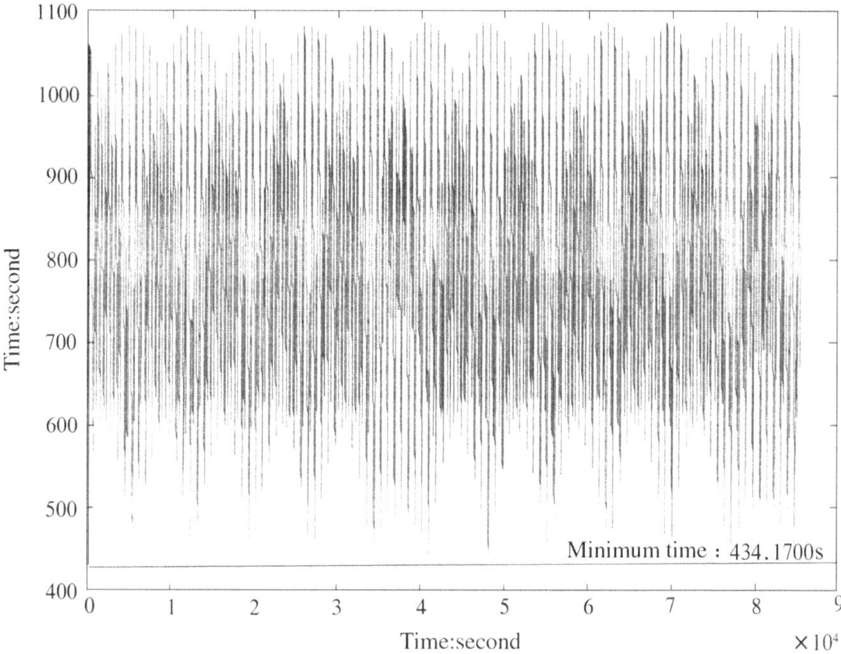

Fig. 2.14 The link duration time of LEO satellites over Beijing station

The simulation results in Fig. 2.14 show that the duration time of the access link is more than 434.17 s, which satisfies the requirement of goal (5), that the access link duration time of any users in any time will last no less than 5 min.

2.4 Analysis of the Constellation Parameters and Their Effect on Routing

The multilayered satellite constellation designed in Sect. 2.2.5 is proved to meet the target (1)–(8) through theoretical analysis and simulation. In order to thoroughly study the needs of constellation design and the implementation of a routing policy, it is necessary to briefly introduce the routing parameters of the constellation.

The routing and switching indicators of the LEO and MEO layers of the multilayered constellation are shown in Table 2.4. Since a GEO layer satellite is usually used as a monitor and backup router in a multilayered constellation, its parameters are not listed here.

The processing capacity of the switching system shown above means the CPU processing power of the onboard router measured by the data packets forwarding rate. The buffer size of the switching system means the buffer size of the onboard router. The above indicators will be different in different constellation systems.

Table 2.4 LEO/MEO satellite routing and switching indicators

	Bandwidth requirements of inter-satellite transceiver (MB/s)	Bandwidth requirements of satellite-ground transceiver (MB/s)	Processing capacity of switching system (kp/s)	Buffer size of switching system (kB)
LEO	≥491	≥380	≥372	≥4.8
MEO	≥191	≥92	≥40	≥0.8

Table 2.5 Service performance of the multilayered constellation system

	Average response time of service (S)	End to end delay (ms)	Delay jitter (s)	IP processing delay (s)
FTP services	41.27	–	–	–
Video services	–	28	1×10^{-3}	–
Voice services	–	19	1×10^{-3}	–
LEO	–	–	–	4×10^{-3}
MEO	–	–	–	1.9×10^{-3}

Under the indicators described in Table 2.4, the performance of FTP services, video services, and voice services provided by LEO and MEO satellites in the multilayered satellite constellation is shown in Table 2.5.

The data in Table 2.5 is based on the indicators in Table 2.4. The Beijing station is used as the receiving station. The ground base station at the same latitude as Beijing and 30° longitude east of from Beijing is used as the sending station. The time of business arrival conforms to the Poisson distribution, and the traffic of business conforms to the simulation data of the terrestrial network in different periods of time. From Tables 2.4 and 2.5, we can see that the requirement of comprehensive indicators of MEO satellites is relatively low, even lower than that of LEO satellites, and the FTP services mainly depend on MEO routing. That means MEO satellites have great potential for the rapid growth of FTP services as well as video and voice services. From the above table, we can see that LEO satellites also work well when their comprehensive performance conforms to the indicators in Table 2.4.

2.5 Summary

This chapter begins with a brief introduction of constellation design and types of constellations, and an analysis of the advantages and disadvantages of different constellations. Then several design goals are proposed to satisfy the QoS requirements of the users. The constellation design is based on these design goals. The effects of layers of a constellation, orbit types, the orbital altitude, the number of satellites, and the arrangement of satellites on the performance of a constellation are all analyzed theoretically. The constellation is designed according to the analysis and is verified through simulation that the design can meet all the design goals. In conclusion, some basic indicators and some basic business processing abilities of onboard routers are given.

References

1. Wood L (2001) Internet working with satellite constellations. Ph.D. thesis, Guildford, University of Surrey
2. Clarke AC (1945) Extra terrestrial relays: can rocket station give world-wide radio coverage. Wirel World 10:208–305
3. Walker JG (1971) Some circular orbit patterns providing continuous whole earth coverage. J Br Interplanet Soc 24:369–384
4. Walker JG (1982) Comments on rosette constellations of earth satellites. IEEE Trans Aerosp Electron Syst 18(4):723–724
5. Walker JG (1984) Satellite constellations. J Br Interplanet Soc 37:559–571
6. Ballard AH (1980) Rosette constellation of earth satellites. IEEE Trans Aerosp Electron Syst 16(5):656–673
7. Rider L (1985) Optimized polar orbit constellations for redundant earth coverage. J Astronaut Sci 33:147–161
8. Rider L (1986) Analytic design of satellite constellations for zonal earth coverage using inclined circular orbit. J Astronaut Sci 34(1):31–64
9. Hanson JM, Higgins WB (1992) Designing good geosynchronous constellations. J Astronaut Sci 38(2):143–159
10. Hanson JM, Evans MJ, Turner RE (1992) Designing good partial coverage satellite constellations. J Astronaut Sci 40(2):215–239
11. Keller H, Salzwedel H, Schorcht G et al (1998) Geometric aspects of polar and near polar circular orbits for the use of ISLs for global communication. In: Proceedings of IEEE 48th vehicular technology conference (VTC'98), Ottawa, vol 5, issue 1, pp 199–203
12. Wu TY (2008) Research of non-geostationary satellite constellation design and inter satellite links. Ph.D. thesis, Chengdu, University of Electronic Science and Technology of China
13. Lee J, Kang S (2000) Satellite over satellite (SOS) network: a novel architecture for satellite network. In: Proceedings of IEEE INFOCOM'2000, Tel-Aviv Israel, vol 3, issue 1, pp 315–321
14. Akyildiz IF, Ekici E, Bender MD (2002) MLSR: a novel routing algorithm for multi-layered satellite IP networks. IEEE/ACM Trans Netw 3:411–424
15. Hu JH, Li T, Wu SQ (2000) Routing of an LEO&MEO double layer mobile satellite communication system. Chin J Electron 28(4):31–35 (in Chinese)
16. Akyildiz IF, Ekici E, Yue G (2003) A distributed multicast routing scheme for multi-layered satellite IP networks. Wirel Netw 9(5):535–544
17. Yuan Z, Zhang J, Liu ZK (2005) A simplified routing and simulating scheme for the LEO/MEO two-layered satellite network. In: Proceedings of the 2005 workshop on high performance switching and routing (HPSR'2005), Hong Kong, vol 5. pp 525–529
18. Pennoni G (1994) JOCOS: 6+1 satellites for global mobile communications. In: Proceedings of IEEE global communication conference (GLOBECOM'1994), San Francisco, vol 3. pp 1369–1374

19. Wu FG, Sun FC, Sun ZQ et al (2005) Performance analysis of a double-layered satellite network. J Comput Res Dev 42(2):259–265 (in Chinese)
20. Leopold RJ, Miller A (1993) The Iridium communication system. IEEE Potentials 12:6–9
21. Chedia L, Smith K, Titzer G (1999) Satellite PCN—The ICO system. Intern J Satell Commun 17:273–289
22. Sturza MA (1995) Architecture of Teledesic satellite system. In: Proceedings of the 4th international mobile satellite conference (IMSC'95), Ottawa, pp 212–218
23. Wiedeman RA, Viterbi AJ (1993) The global star mobile satellite system for worldwide personal communications. In: Proceedings of the 3rd international mobile satellite conference (IMSC'93), Pasadena, pp 291–296

Chapter 3
Satellite Network Routing Strategies

3.1 Introduction

The satellite network routing strategy is the basis for the application of satellite network routing algorithms and traffic engineering. It is mainly related to the solution of the problem of time variance of satellite network topologies. Routing strategies are an important part of satellite network routing, for it can directly affect the performance of routing algorithms and traffic engineering. Two main satellite network routing strategies, the virtual node strategy and the virtual topology strategy, are introduced in the former Sect. 1.3.3. This chapter will make a brief review and an in-depth discussion of these two strategies, and then present a novel satellite network routing strategy for a multilayered satellite network. Satellite network systems are inherently suitable for broadcasting and can provide broadband access for the end users that are not suitable for cable access to the Internet. Compared with single-layered satellite networks, multilayered satellite networks are more robust and survivable, and the routing strategies fit for multi-layered satellite networks are more diverse. Many satellite constellation systems provide inter-satellite communication ISLs, and these ISLs can be used to transmit network control and data signals [1]. The following researches are based on the constellation system with inter-satellite ISLs.

The rapid movement of non-geostationary satellites leads to a continuous change of the topology of a multilayered constellation system. Meanwhile, the connection states of the ISLs change with the change of the distance and azimuth between two endpoint satellites. Hence, developing routing algorithms for this type of dynamically changing topology is quite challenging. The dynamically changing topology of a satellite network is the foremost problem to be solved by routing algorithms. Two intuitive ideas are to shield the dynamic topology, and to transform the continuously changing topology into a series of discrete static topologies. The virtual node strategy is derived from the first idea while the virtual topology strategy is derived from the latter.

With the rapid growth of Internet applications, satellite networks are required to transmit IP traffic [2]. A lot of LEO satellite network routing protocols based on IP

F. Long, *Satellite Network Robust QoS-aware Routing*,

DOI: 10.1007/978-3-642-54353-1_3,

were proposed. The DRA routing algorithm [3] proposed by Ekici E et al. is the most typical virtual node strategy routing algorithm based on IP. The DRA algorithm consists of three parts: a virtual node routing strategy based on the LEO Walker constellation, a two-stage routing algorithm, and a congestion control algorithm. The original intention of DRA is to solve the problem of path switching [7] of the connection oriented ATM onboard routing architecture [4–6], and the long end-to-end routing delay of the darting algorithm [8]. The DRA algorithm improves the datagram routing algorithm [9]. It uses propagation delay as the optimization objective and routes in a distributed manner.

In the virtual node routing strategy proposed in [3], the geographic position of Satellite S is identified by a tuple $< lon_s, lat_s >$, in which lon_s is the longitude of S and lat_s is the latitude of S. Assuming that the surface of the Earth is covered by such logical positions according to the geographic positions of satellites. These logical positions are fixed and each is bound by the satellite nearest to it. However, the geographic position of Satellite S is not permanently bound to its logical position. When Satellite S leaves a geographic position, its current logical position will be replaced by the successor satellite in the same track. The logical position of Satellite S is identified by tuple $<p, s>$, in which $p = 0, \ldots, N-1$ is the track number, and $s = 0, \ldots, M-1$ is the satellite number. A logical position is simply deemed as a hop in the routing algorithm, which can shield the movement of the satellites.

Although the virtual node strategy proposed in [3] is easy to be implemented and has a little protocol overhead, its disadvantages are obvious too. First, constellation topology information is not included in the routing strategy, which makes a long recovery time when a satellite fails. Second, the application scope is narrow. It is neither not suitable for multilayered satellite constellation structure, nor needs large modification before using in a nonpolar orbit LEO constellation. Last, only the propagation delay included in the routing selection can not reflect the total delay of the network packet transmission, the total delay of packet transmission should include propagation delay, queuing delay, and processing delay.

The main idea of the virtual topology strategy is to convert the continuous dynamic satellite network topology into a series of fixed topologies. It was first proposed by Werner in [10], and was developed in [11]. The method in [11] converts the dynamic topology into a series of fixed topologies with equal time length. The optimized link allocation strategy is applied to each fixed topology. After the Satellite over Satellite (SoS) structure was proposed in [12], the multilayered satellite network has gradually become the research focus in the field of satellite network. In order to reduce the routing hops and resource consumption, the idea of using MEO satellites to carry the long distance-dependent traffic was first proposed in [12]. A routing strategy based on the SoS structure was proposed in [13]. However, it assumed that there are no ISLs between LEO satellites, all the network routing function is afforded by MEO satellites.

The MLSR protocol proposed in [14] by Akyildiz is a classic virtual topology strategy. MLSR is based on an LEO/MEO/GEO triple-layered satellite network. It expands the application of the virtual topology strategy to a multilayered satellite constellation for the first time. The MLSR collects the delay information periodically, and calculates the routing based on it. In the MLSR protocol, several LEO satellites are organized into a team with an MEO satellite as its team leader. The topology of a LEO group is invisible to the other satellites except its team leader, for the group is abstracted into a meta-node in the routing calculation. The routing calculation is carried out by GEO satellites, and the satellite network topology graph composed by meta-nodes is used for the routing to reduce the computing work load. The calculated routing table is distributed by MEO satellites to its group members after optimization then. The idea of low layer satellite grouping is first implemented in MLSR, and the virtual topology is divided by the grouping situation. However, the MEO satellites are sparsely distributed in many cases. In these cases, the LEO satellites may not have their group leaders; hence the MLSR can not be implemented. Moreover, the MLSR depends on the periodically routing information collection to solve the congestion problem, and lacks a rapid response mechanism to handle it.

After the MLSR was proposed, the satellite grouping and routing protocol (SGRP) was proposed by Chen et al. in 2002. SGRP was first proposed in [15], and improved in [16], and finally described in detail in [17]. SGRP inherited the thinking of satellite grouping in MLSR. It is applied in the LEO/MEO double-layered satellite constellation, and takes full advantage of the collaborative relationship between the LEO layer and MEO layer satellites. The MEO satellites are selected as the manager of the LEO satellites. The main idea of SGRP is to transmit the datagram along the shortest delay path, and transfer the task of routing calculation to the MEO layer satellites. Same as MLSR, SGRP also divide the LEO satellites into several groups; the LEO satellites within the same footprint of an MEO satellite are classified as a group. When an LEO satellite leaves the footprint of an MEO satellite, the group membership will change. In SGRP, each change in group membership is deemed as a topology change, and each change will generate a new snapshot. The satellite network topology can be deemed as fixed in each snapshot. Each MEO satellite is the group manager of the LEO group in its footprint. The group manager is in charge of collecting and exchanging the routing information of the LEO layer satellites and calculating the routing table for the LEO satellites. The LEO satellites receive the routing tables from the MEO satellites and forward packets according to the routing tables. The greatest advantage of SGRP is that it moves the routing computing task to the more powerful MEO satellites, which efficiently balances the resource consumption of LEO and MEO satellites and extends the life of the whole satellite network system. The SGRP separates the data flow and signal flow physically. As a result, the link congestion cannot delay the transfer of routing information and the routing computation.

Although SGRP has solved the problem of inefficient congestion handling in MLSR, improved the LEO grouping mechanism, and reduced the routing calculation burden of LEO satellites, there are still the following problems to be resolved.

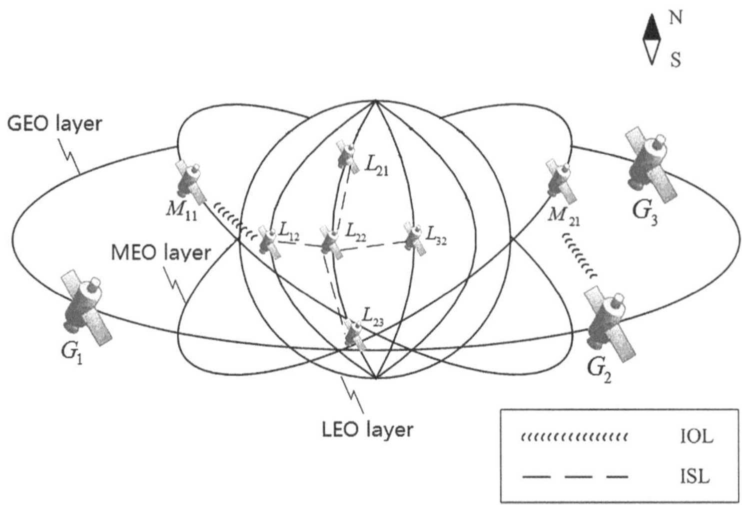

Fig. 3.1 The *Tr* constellation model schematic diagram

(1) The task of routing computation will increase the burden of MEO satellites, thus reduce the life of the entire system, and its robustness and survivability.
(2) The change of LEO grouping is bound to the change of LEO logical position, which will increase the number of snapshots in one system cycle, hence increase the protocol overhead and the system burden.
(3) The number of snapshots is too large as the LEO satellites frequently enter and leave the footprints of MEO satellites, which is not conducive to the application of the routing algorithm.
(4) The extensibility of the routing strategy is not good enough.

In the following part of this chapter, a novel satellite grouping and routing protocol (NSGRP) will be proposed. It combines the advantages of the virtual node strategy and the virtual topology strategy. Through simulations, we know that the NSGRP can solve the problems in MLSR and SGRP mentioned above, and performs well.

3.2 The Components of NSGRP

The *Tr* constellation mentioned in Sect. 2.2.5 is used as our experimental model. The NSGRP is designed to be applied in the *Tr* constellation. Each part of the *Tr* constellation is a component of the NSGRP routing strategy. The components of the *Tr* constellation is shown in Fig. 3.1.

According to the *Tr* constellation shown in Fig. 3.1, the role and function of each component of the NSGRP will be described in detail as follows.

3.2.1 Intersatellite Links

The *Tr* constellation is different from the multilayered constellation that has no interlayer links proposed in [13]. In the *Tr* system, the terrestrial gateway links with the LEO, MEO, and GEO layers. There are interlayer links between the LEO–MEO, LEO–GEO, and MEO–GEO layers. Meanwhile, each layer in *Tr* has its own inner layer links. The *Tr* system has three full-duplex links altogether.

1. ISL: Satellites s and d within the same layer communicate with each other through ISLs identified by $\mathrm{ISL}_{s \to d}$ or $\mathrm{ISL}_{d \to s}$. An LEO satellite keeps four ISLs, two of which connect the neighboring LEO satellites in the same orbit while the other two connect the neighboring LEO satellites in the two neighboring orbits. The LEO satellites on both sides of the reverse seam keep only three ISLs. The MEO satellites keep two permanent ISLs. Each of them connects a neighboring MEO satellite in the same orbit plane. The temporary ISL is established between the neighboring MEO satellites in the two neighboring orbits when they move into the adjacent area of the intersection of the two orbits. The information will be exchanged in that case. The GEO satellites only keep two ISLs to connect with its two neighboring GEO satellites.
2. IOL: The satellites within different layers communicate with each other through IOL. If LEO satellite s is located in the footprint of MEO satellite d, they are connected by IOL. Similarly, If MEO satellite s is located in the footprint of GEO satellite d, there exists a IOL between them too. These two situations are both identified as $\mathrm{IOL}_{s \to d}$ or $\mathrm{IOL}_{d \to s}$.
3. UDL: The terrestrial gateway communicates with LEO, MEO, or GEO satellites through UDL. The UDLs connecting the LEO/MEO/GEO satellites and the terrestrial gateway are identified as $\mathrm{UDL}_{s \to G}$ or $\mathrm{UDL}_{G \to s}$.

3.2.2 GEO Satellites

The GEO satellites can fully cover the MEO and LEO layer satellites and the Earth surface. This character makes the GEO satellites the backup and monitor satellites in NSGRP. GEO satellites are responsible for monitoring the operation of the satellites in the LEO and MEO layers. Once receiving the report of LEO, MEO failure or ISL, IOL congestion, it will immediately flood the entire GEO layer and inform the terrestrial gateway in its coverage area. The GEO satellites can also be used as the bypass router for the LEO/MEO satellites, and share some traffic with low priority when some LEO/MEO satellites have a high traffic load or even fail to operate. Besides, the GEO satellites are in charge of recalculating the snapshots and spreading them to the LEO/MEO satellites and terrestrial gateways when LEO/MEO satellites fail or the link is congested. Although NSGRP is used in the *Tr* constellation, it can be extended to other types of constellations too. When the lower satellite layers of the constellation are not a polar orbit (LEO layer) and 45° inclined orbit (MEO layer) constellation as mentioned in *Tr*, GEO satellites can be used as the group managers of lower layer satellites, and play the same role as the MEO satellites in SGRP. There are $N_G = 3$ GEO satellites, identified as GS_i, $i = 1,\ldots,N_G$.

3.2.3 MEO Satellites

The LEO satellites are divided into several groups according to the footprints of the MEO satellites. Similar to SGRP, the MEO satellites are also used as the group managers of LEO satellites. The system cycle of the whole satellite network system is divided into several snapshots according to the grouping status under each MEO manager. The grouping of the whole system is fixed in each snapshot. At the beginning of each snapshot, each MEO satellite collects the LEO link state information in its group, and reports it together with the state information of the ISLs which are connected to itself to the GEO satellite who is in charge of it. The link state of the LEO layer will also be broadcasted in the MEO layer, so that the whole MEO layer can know the topology and link state of the LEO/MEO layer. The MEO satellites will calculate the routing table of its LEO group members according to the topology and link state information of the LEO/MEO layer it owns. At the same time, it reports the topology and link state information to the terrestrial gateway. When receiving the failure or link congestion report of the LEO layer, the MEO satellite that is in charge will report to the terrestrial gateway in its footprint and the manager GEO satellite too. The total number of satellites of the *Tr* constellation is $N_M \times M_M$, where $N_M = 2$ is the number of orbit planes in the MEO layer and $M_M = 5$ is the number of satellites in each orbit plane. The MEO satellites are identified as $MS_{i,j}$, in which $i = 1,\ldots,N_M, j = 1,\ldots,M_M$.

3.2.4 LEO Satellites

In NSGRP, the LEO satellites also use the logical locations in the virtual node strategy to shield their mobility as opposed to the ground. As a result, an LEO satellite can be identified by two IDs, i.e., the physical serial number and its logical location. The physical serial numbers are expressed as $PLS_{i,j}$, $i = 1,\ldots,N_L$, $j = 1,\ldots,M_L$, where $N_L = 6$ is the number of orbit planes in the LEO layer, and $M_L = 11$ is the number of satellites in each orbit plane in the LEO layer. The physical serial number is decided by the initial launch position of the entire constellation, which is similar to the MAC address of the host in the terrestrial network. The logical locations are expressed as $LLS_{i,j}$, $i = 1,\ldots,N_L$, $j = 1,\ldots,M_L$, which is fixed to the ground. Each logical location is obtained through the nearest LEO satellite, which is similar to the IP address in the terrestrial network. When an LEO satellite leaves a certain logical location, it will forward all the routing information related to this logical location to its successor satellite. When the logical location changes but the grouping does not change, it does not produce a new snapshot. At the beginning of each snapshot, each LEO satellite needs to submit all the link state information that is related to it to its MEO group manager. The LEO satellites will forward data packets for ground users in each snapshot according to the routing table sent by its group manager. It needs to report immediately to its MEO manager when the failure of its adjacent satellite or its output/input links is found.

3.2.5 Terrestrial Gateways

Different from the SGRP strategy, the terrestrial gateway needs to take certain routing computation task in NSGRP. The NSGRP classifies the routing into ordinary routing and QoS routing. The ordinary routing uses delay as the optimization objective while the QoS routing needs to satisfy the users' QoS requirements. The ordinary routing calculation is carried out by the MEO satellites, and the QoS routing is carried out by the terrestrial gateway. The terrestrial gateway is identified as TG_i. The detailed functions are as follows:

1. Snapshot calculation: In the initial state of the constellation, it will calculate the snapshots using the NSGRP strategy and then upload result to the GEO satellite for distribution.
2. QoS routing: Upon receiving the QoS routing requirements of the users, it will calculate all the routes that meet the users' QoS requirements in a source routing way.
3. Address translation: It is responsible for address translations between the terrestrial AS and the space-based network. Assuming that data packets need to be transferred from terrestrial gateway TG_1 to terrestrial gateway TG_2 through the space-based network, TG_1 will upload them through $UDL_{TG_1 \to LLS_{n_1,m_1}}$ to LLS_{n_1,m_1}, according to its logical location $< n_1, m_1 >$ when receiving the forwarding request, satellite LLS_{n_1,m_1} will then route the packets to the satellite LLS_{n_2,m_2} that is bound to the logical location $< n_2, m_2 >$. Finally, the packets will be transmitted through $UDL_{LLS_{n_2,m_2} \to TG_2}$ to terrestrial gateway TG_2. In this process, the task of translating the terrestrial address of TG_2 to the logical address that can be recognized by satellites is carried out through gateway TG_1. r.

The above-mentioned components carry parts of the NSGRP function modules respectively, and serve the terrestrial business in common. The organization of LEO/MEO/GEO layers is shown in Fig. 3.2. An LEO satellite may be covered by several MEO satellites. In this case, it will join the group whose MEO leader has the longest coverage time. Similarly, an MEO satellite will choose the GEO satellite that has the

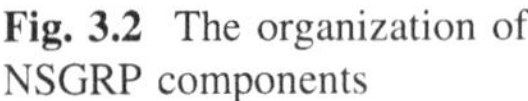

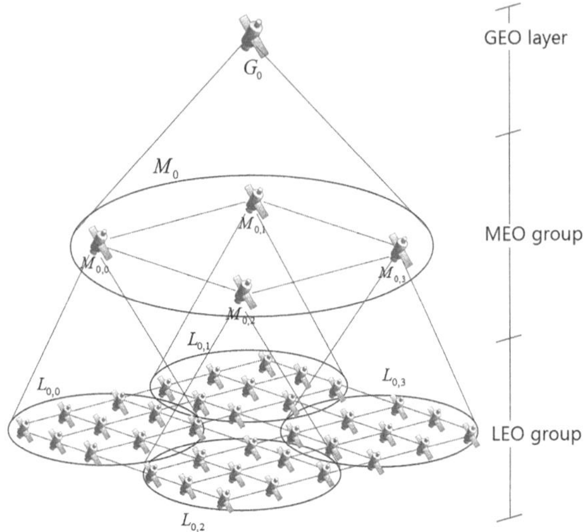

Fig. 3.2 The organization of NSGRP components

longest coverage time as its manager when it is covered by more than one GEO satellite. The implementation of NSGRP will be discussed in Sect. 3.3.

3.3 The Implementation of NSGRP

3.3.1 Concepts and Definitions

Before describing the implementation of NSGRP, it is necessary to define the concepts in the strategy. The concepts that appear in NSGRP are defined as follows:

Definition 3.1 LEO group: An LEO group $L_{i,j}$ is the set of LEO satellites that are located in the footprint of MEO satellite M_{ij}, and M_{ij} is chosen as their manager according to the longest coverage time rule. $L_{i,j} = \{L_{i,j,k} | k = 0, \ldots, S(L_{i,j}) - 1\}$, in which $L_{i,j,k}$ represents the kth LEO satellite in group $L_{i,j}$, and function $S(L_{i,j})$ returns the number of satellites of group $L_{i,j}$.

Definition 3.2 MEO group: An MEO group M_i is the set of MEO satellites that are located in the footprint of GEO satellite G_i, and G_i is chosen as their manager according to the longest coverage time rule. $M_i = \{M_{i,k} | k = 0, \ldots, S(M_i) - 1\}$, in which $M_{i,k}$ represents the kth MEO satellite in group M_i, and function $S(M_i)$ returns the number of satellites of group M_i.

Definition 3.3 Group manager: In NSGRP, there is no affiliation between LEO satellites and GEO satellites. If a high layer satellite H_i has the longest time to cover a low layer satellite D_i in the tth snapshot, it is called the group manager of satellite D_i, and is denoted as $GM_t(D_i) = H_i$.

Definition 3.4 Comprehensive link information report: The comprehensive link information report $LMR(X)$ of satellite node X is a set of tuples $\{Y, C(l_{X\to Y})\}$, in which Y is a satellite node that connects to X with link $l_{X\to Y}$, and $C(l_{X\to Y})$ is a comprehensive link report including link delay, residual bandwidth and packet loss rate.

$$C(l_{X\to Y}) = \begin{cases} (td(X,Y), rb(X,Y), pl(X,Y)), & \exists l_{X\to Y} \\ (\infty,\infty,\infty), & \text{otherwise} \end{cases}$$

$td(X, Y)$ is the end-to-end delay from X to Y, including the queuing delay and processing delay in node X, and the propagation delay of ISL $l_{X\to Y}$; $rb(X, Y)$ is the residual bandwidth of the ISL $l_{X\to Y}$; $pl(X, Y)$ is the loss rate of the ISL $l_{X\to Y}$.

Definition 3.5 List of group members: Each MEO and GEO satellite has a list of its group members in each snapshot. The lists of group members in snapshot t of the MEO satellite $M_{i,k}$ and the GEO satellite G_i are denoted as $GL_t(M_{i,k})$ and $GL_t(G_i)$ respectively and defined as follows:

$$GL_t(M_{i,k}) = \{L_{i,k,x} | GM_t(L_{i,k,x}) = M_{i,k}\}$$

$$GL_t(G_i) = \{M_{i,x} | GM_t(M_{i,x}) = G_i\}.$$

Definition 3.6 Ascending/descending point of MEO orbits: In the *Tr* constellation, there are two cross-points in the MEO layer. If the MEO satellites cross the point from south to north, it is called an ascending point of the MEO orbits; otherwise, it is a descending point.

In the *Tr* constellation, there are two MEO orbits each with five satellites. This satellite arrangement makes at least one satellite in another orbit visible to the satellite near the ascending/descending point (proof omitted). These two satellites will then establish inter-orbit ISL at this time and exchange network topology information.

Definition 3.7 The shortest delay path: $P_{X\to Y}$ is defined as the satellites serial in the shortest delay path from the source satellite X to the destination satellite Y.

Definition 3.8 Routing table of the MEO layer: The MEO layer satellites have two roles in NSGRP, leaders of the LEO group and ordinary router. As a result, the routing table of the MEO satellite $M_{i,j}$ is composed of two parts: the original routing table of the LEO layer and the forwarding table for the MEO layer. The routing table of the LEO layer *LOT* records the paths from the satellites in $L_{i,j}$ to all the terminal satellites, and is defined as follows:

$LOT(M_{i,j}|X) = \{<X, Y, P_{X\to Y}> | Y \in \text{LEO layer of the } Tr \text{ constellation}\}$, in which $X \in L_{i,j}$.

The forwarding table of MEO layer *MRT* records the next hops from satellite $M_{i,j}$ to all the terminal satellites, and is defined as follows:

$MRT(M_{i,j}) = \{<Y, fe(P_{M_{i,j}\to Y})> | Y \in \text{LEO layer of the } Tr \text{ constellation}\}$, in which $fe(P_{X\to Y})$ returns the first element of $P_{X\to Y}$.

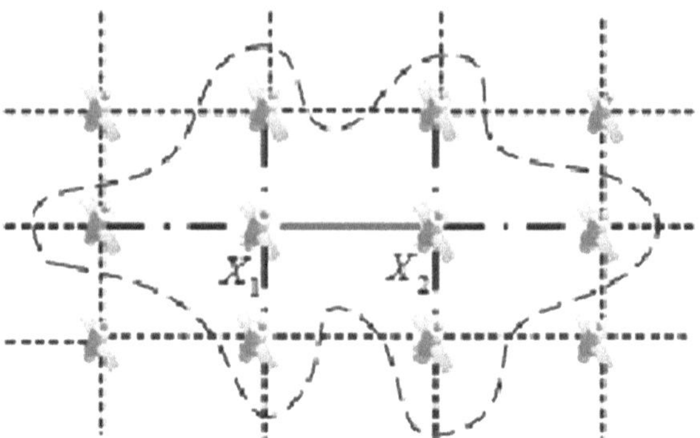

Fig. 3.3 The congestion area of $\mathrm{ISL}_{X_1 \to X_2}$

Definition 3.9 Routing table of the LEO layer: The routing table of the LEO layer LRT is calculated and sent by the MEO manager, and is defined as follows:

$LRT(L_{i,j,k}) = ft(LOT(M_{i,j}|L_{i,j,k}))$, in which $ft(x)$ returns the first three rows of x.

Obviously, LRT is a next hop forwarding table.

When a LEO satellite $PLS_{i,j}$, $LLS_{i,j}$ leaves its logical location $<i,j>$, if it is not the time for snapshot switching, then it only needs to transfer the routing table $LRT_{i,j}$ to the satellite $PLS_{i,j+1}$, and the identifier of satellite $PLS_{i,j}$ becomes $\mathrm{PLS}_{i,j}$, $LLS_{i,j-1}$.

Definition 3.10 Link congestion warning: When a link utility of an $\mathrm{ISL}_{X \to Y} \geq 95\,\%$, the satellite X will report to its manager $alert(\mathrm{ISL}_{X \to Y})$, which is called link congestion warning. Its function is to remind the upper layer manager to recalculate the route.

Definition 3.11 Satellite failure report: When satellite X finds its adjacent satellite Y failing, it will report to its manager $alert(Y)$, which is called satellite failure report. Its function is to remind the manager to recalculate the route and then to inform the ground base station for subsequent processing.

Definition 3.12 Congestion area: When $\mathrm{ISL}_{X_1 \to X_2}$ is congested, it will block links in a certain area, which is hence called the congestion area of this ISL. The congestion area $CGA(\mathrm{ISL}_{X_1 \to X_2})$ is defined as

$$CGA(\mathrm{ISL}_{X_1 \to X_2}) = \cup\{\mathrm{ISL}_{Y_1 \to Y_2} | Y_1 \neq Y_2, \quad Y_1 \text{ or } Y_2 = X_i, \quad i = 1,2\}.$$

The congestion area of $\mathrm{ISL}_{X_1 \to X_2}$ is shown in Fig. 3.3. The solid line represents the congested link, the dash dot line represents blocked links, and the dashed curve represents the congestion area.

3.3.2 Implementation of NSGRP

The design goal of NSGRP is to shield the mobility of satellites. The NSGRP will forward the packets along the shortest delay path while providing QoS routing for the users in need. After introducing the components and basic concepts of NSGRP, this section will discuss the implementation of NSGRP in six parts:

(1) Division of snapshots and its improvement in system cycle.
(2) Production of *LMR*.
(3) Transmission of *LMR*.
(4) Calculation of ordinary routing table.
(5) Congestion avoidance.
(6) Treatment of node failure.

3.3.2.1 Division of Snapshots

To apply normally the routing algorithm under the satellite network environment, the first problem to be solved is the high-speed change of the satellite network topology. The solution of the time-varying problem of the satellite network topology depends on the choice of a topology fixing strategy. In the NSGRP, the virtual node strategy is adopted in the LEO layer, and the improved virtual topology grouping strategy is used in MEO and GEO layers. Since LEO/MEO satellites carry out most of the routing and data transmission tasks, the affiliation of LEO satellites to MEO satellites is used to divide the topology snapshots in the virtual topology grouping strategy. The division of topology snapshots affects the performance of the virtual topology grouping strategy and even the performance of the entire system, hence playing an important role in the NSGRP. As the implementation of the virtual node strategy in the LEO layer was already discussed above, there is no need to discuss it here. The implementation of topology snapshot division according to the affiliation of LEO satellites to MEO satellites is described as follows:

The LEO/MEO constellation is extracted from *Tr* as shown in Fig. 3.4, where the shadow circle presents the coverage area of MEO satellites over LEO satellites, and the hollow circle presents the coverage area of LEO satellites over the Earth. The LEO satellite which is covered by more than one MEO satellite will choose the MEO satellite that has the longest coverage time to join in as mentioned above. The ground base station in a high latitude area, which is covered by more than one LEO satellite, can choose the LEO satellite that has the longest serving time as its interface satellite to reduce access switching.

In the NSGRP, each LEO satellite chooses an MEO satellite that has the longest coverage time as its current group leader. When the LEO–MEO elevation angle of an LEO satellite is less than the critical angle, the LEO satellite will leave its current group and choose another MEO satellite to join in; a new snapshot is hence generated. The grouping is fixed in each snapshot, thereby the topology of the whole network can be deemed as fixed.

In order to calculate the snapshot division of the satellite network, the exact position of each satellite at time t should be known. The grouping of LEO satellites can be uniquely determined by the position information. In this section, the coordinate system is established as shown in Fig. 3.5, in which the polar axis of the Earth is the z axis with north as its positive direction, the directed line from the

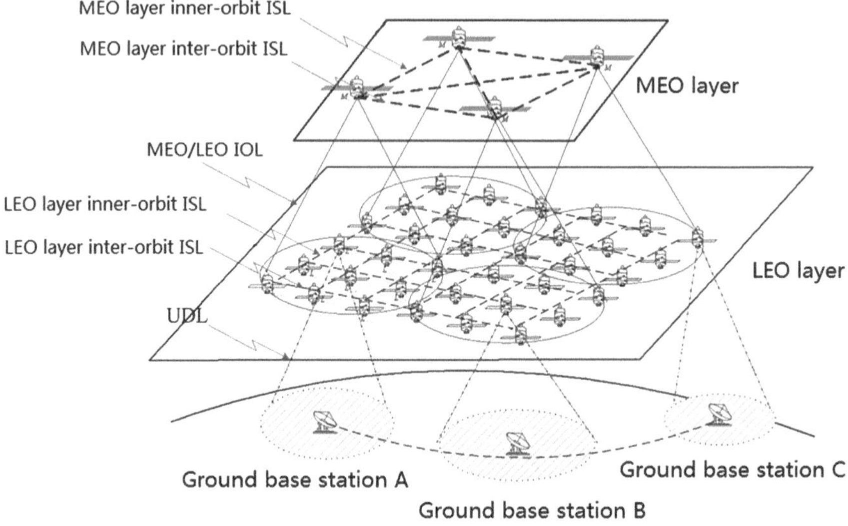

Fig. 3.4 LEO/MEO layers in *Tr* constellation

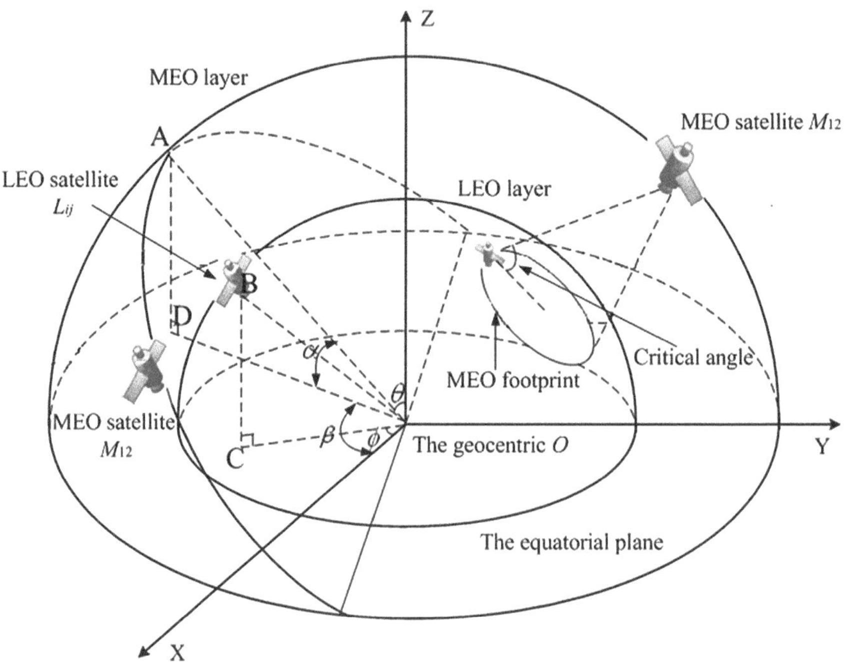

Fig. 3.5 Coordinate of LEO/MEO layers

geocentric to the point (N:0°,E: 0°) is the positive direction of x axis. The directed line from the geocentric to the point (N:0°,E: 90°) is the positive direction of y axis.

In Fig. 3.5, $\alpha = \angle AOD$ is the included angle between line segment AO and the equatorial plane (assume that point A is a position of MEO satellite M_{21}); $\beta = \angle DOX$ is the included angle between line segment DO and the positive direction of X axis, in which D is the projection of point A in the equatorial plane; $\theta = \angle BOZ$ is the included angle between line segment BO and the positive direction of Z axis, in which B is a position of LEO satellite L_{ij}; $\phi = \angle COX$ is the included angle between line segment CO and the positive direction of X axis, in which C is the projection of point A in the equatorial plane. The minimum elevation angle of LEO–MEO determines the coverage area of MEO satellite to LEO layer.

According to Fig. 3.5, the coordinate of LEO satellite in time t is

$$\begin{cases} X_L = R_L \sin(\theta + \omega_L t) \cos\phi \\ Y_L = R_L \sin(\theta + \omega_L t) \sin\phi \\ Z_L = R_L \cos(\theta + \omega_L t) \end{cases} \tag{3.1}$$

in which

$$\begin{cases} \theta = (i-1)\dfrac{2\pi}{mn} + (j-1)\dfrac{2\pi}{n} \\ \phi = (i-1)\dfrac{\pi}{m} - \dfrac{\pi}{2} \end{cases} \tag{3.2}$$

R_L is the orbital radius of the LEO satellite; w_L is the angular velocity of the LEO satellite; (i, j) is the ID of the LEO satellite, in which i is the track number and j is the satellite number in each track; m is the amount of orbits in the LEO layer; n is the amount of satellite in each LEO orbit.

Similarly, the coordinate of MEO satellite in time t is

$$\begin{cases} X_M = R_M \sin\left(\omega_M t + (l-1)\dfrac{2\pi}{k}\right)\cos\left(\beta + \dfrac{\pi}{2}\right) + R_M \cos\left(\omega_M t + (l-1)\dfrac{2\pi}{k}\right)\sin\left(\beta + \dfrac{\pi}{2}\right)\cos\alpha \\ Y_M = R_M \sin\left(\omega_M t + (l-1)\dfrac{2\pi}{k}\right)\sin\left(\beta + \dfrac{\pi}{2}\right) - R_M \cos\left(\omega_M t + (l-1)\dfrac{2\pi}{k}\right)\cos\left(\beta + \dfrac{\pi}{2}\right)\cos\alpha \\ Z_M = R_M \cos\left(\omega_M t + (l-1)\dfrac{2\pi}{k}\right)\sin\alpha \end{cases} \tag{3.3}$$

in which R_M is the orbital radius of the MEO satellite; w_M is the angular velocity of the MEO satellite; l is the track number of the MEO satellite and k is the satellite number in each track.

If the coordinate of LEO and MEO satellites in any time are known, the changing moment of the LEO group, which is also the partition point of snapshots, can be calculated according to the least elevation angle between LEO and MEO satellite. The least elevation of LEO–MEO is shown in Fig. 3.6. We assume that the least elevation angle is 10° here.

According to Fig. 3.6 and the law of sines, we have

$$\frac{R_M}{\sin(90^\circ + \varepsilon_{\min})} = \frac{R_L}{\sin(180^\circ - 90^\circ - \varepsilon_{\min} - c)} \tag{3.4}$$

According to Eq. (3.4), we have

$$c = 80^\circ - \arcsin\left(\frac{R_L}{R_M}\sin 100^\circ\right) \tag{3.5}$$

Using the law of cosines, we have:

$$R_L^2 + R_M^2 - 2R_M R_L \cos c = d_{\text{LEO-MEO}}^2 \tag{3.6}$$

in which

$$d_{\text{LEO-MEO}} = \sqrt{(X_L - X_M)^2 - (Y_L - Y_M)^2 - (Z_L - Z_M)^2} \tag{3.7}$$

Substituting (3.1), (3.3), (3.5), (3.7) into (3.7), we have the equation for the critical angle. Solving t from the equation, we can get the group switching time of LEO satellites, which is also the division of snapshots. The solutions of the equation are identified as t_i, $i = 1, 2, \ldots, p$, where p is the number of snapshots in the system cycle, t_i is the snapshot switching time.

Since LEO/MEO satellites carry out most of the routing and data transmission tasks, the virtual node strategy is used to shield the movement of the LEO layer. For ease of computing, the least common multiple of running cycles of LEO layered satellites and MEO layered satellites is used as the system cycle of the *Tr* constellation and the snapshot cycle. The system cycle is hence 12 h, and the division of snapshots in the system cycle is shown in Fig. 3.7.

Figure 3.7 shows the division of snapshots using the virtual topology strategy in SGRP. The system cycle of *Tr* is divided into 2,670 snapshots, the average length of the snapshots is 16.16 s; the minimum length of the snapshots is only 0.0075 s; and the maximum length of the snapshots is 134.92 s. Obviously, the snapshot partition method in SGRP makes the minimum length of the snapshot in the system cycle too short. Moreover, the average length of the snapshots is not long enough. The frequent snapshot switching brings a large routing protocol overhead, which is not suitable for the application of routing algorithms within the snapshots.

The virtual topology grouping strategy in SGRP is fit for the constellations that have a relatively small number of LEO satellites and a large height difference between the LEO and MEO orbits. This is because a smaller number of LEO satellites will reduce the chance of group changing of the LEO satellites, and reduce the snapshots in one system cycle consequently. The increase of the MEO satellite orbit height will enlarge its footprint on the LEO layer, which will also reduce the frequency of group changing. However, the reduction of LEO satellites may cause an incomplete coverage of the Earth, and the increase of the MEO orbit

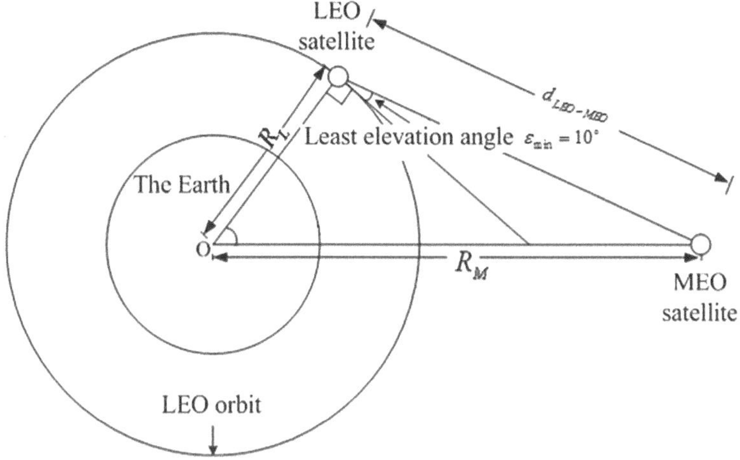

Fig. 3.6 The least elevation angle of Leo–MEO

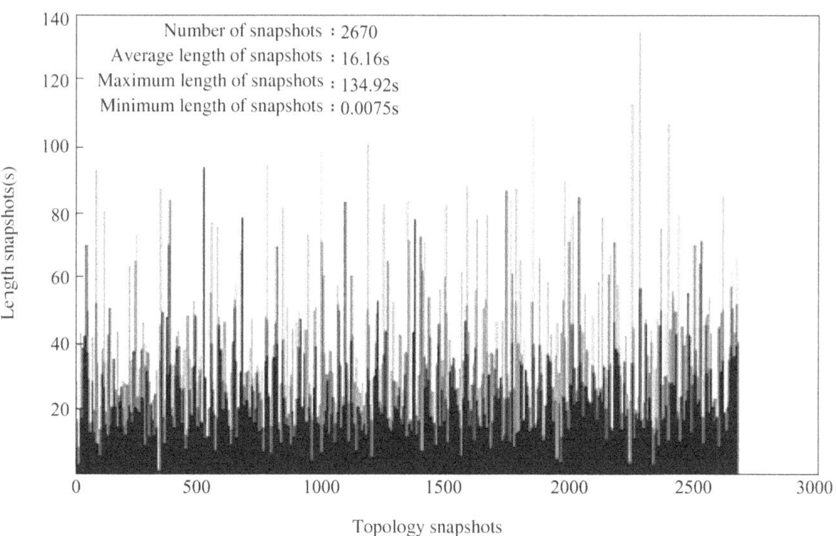

Fig. 3.7 Division of snapshots in SGRP

height will increase the propagation delay and the system cycle. Both of them can affect the performance of the entire satellite network system. To make the virtual topology strategy fit for the situation of a large number of LEO satellites and a moderate height of MEO orbits, the snapshot division method in SGRP must be improved.

3.3.2.2 Improvement of Snapshot Division

In the SGRP, an LEO satellite leave or join an MEO group will generate a new snapshot. Since the *Tr* constellation has 66 LEO satellites, the leaving and joining of an LEO satellite in one system cycle will generate a large number of snapshots. There are two intuitive ideas to reduce the snapshots in the system cycle. One is to re-divide the snapshots, and the other is to merge some snapshots based on the existing division of snapshots. Since it is difficult to re-divide the snapshots under the framework of the virtual topology grouping strategy in an LEO/MEO constellation, the second idea is adopted to improve the snapshot division.

Let's consider the generation of snapshots first. An LEO satellite $L_{i,j,k}$ leaves group $L_{i,j}$ at time t_p when its group manager $GM_{p-1}(L_{i,j,k})$ cannot cover it, and joins a new group $L_{i,j+1}$ whose manager can provide the longest coverage time. This generates a new snapshot $[t_{p-1}, t_p]$. Through simulation we know that the LEO satellite $L_{i,j,k}$ can be covered by both MEO satellites $GM_{p-1}(L_{i,j,k})$ and $GM_p(L_{i,j,k})$ at a certain period of time $[t_p - \Delta, t_p]$ before the switching time t_p. In this case, LEO satellite $L_{i,j,k}$ does not necessarily have to switch at time t_p. In other words, it can switch at any time during the period $[t_p - \varDelta, t_p]$. The period $[t_p - \varDelta, t_p]$ is called the free period of LEO satellite $L_{i,j,k}$. The existence of the free period in the system cycle makes snapshot merging possible.

For ease of understanding, it is assumed that the switching time set based on the method of SGRP is $[t_1, \ldots, t_i, \ldots, t_n]$. The LEO satellite L_i switches at time t_i, and $GM_i(L_i) = M_i$, $GM_{i+1}(L_i) = M_{i+1}$. Let's first consider the partitioned time t_1. If it exists in the free period of LEO satellite L_2, then L_2 can switch at time t_1. Therefore, a snapshot can be merged in this way. Similarly, we can merge all the snapshots that meet the conditions, and a new snapshot division can be obtained this way. The principle of snapshot merging is shown in Fig. 3.8.

T_i shown in Fig. 3.8 is the time division after snapshot merging. The pseudo code of the snapshot merging is shown in Table 3.1.

The snapshot division calculated by Algorithm 3.1 is shown in Fig. 3.9.

As shown in Fig. 3.9, the number of topology snapshots in a system cycle decreases from 2,670 to 322. The average length of snapshots increases from 16.16 to 133.77 s. The shortest length of snapshots increases from 0.0075 to 8 s accordingly, which basically satisfies the needs of the application of the routing algorithms. Intuitively, a higher MEO orbit height leads to a longer free period of LEO satellites, which brings more chances for the merging of snapshots. However, the increase of the MEO orbit height can also lead to a larger RTT and system cycle, which will worsen the network performance and communication quality. The designer should choose the MEO orbit height carefully according to certain design goals.

The quality of snapshot division greatly affects QoS routing, which will be discussed in Chap. 5. It is a symbol of the changing frequency of the network topology. If there is snapshot debris during the snapshot division, some QoS routing algorithms, which will be mentioned later, may not be applicable for they

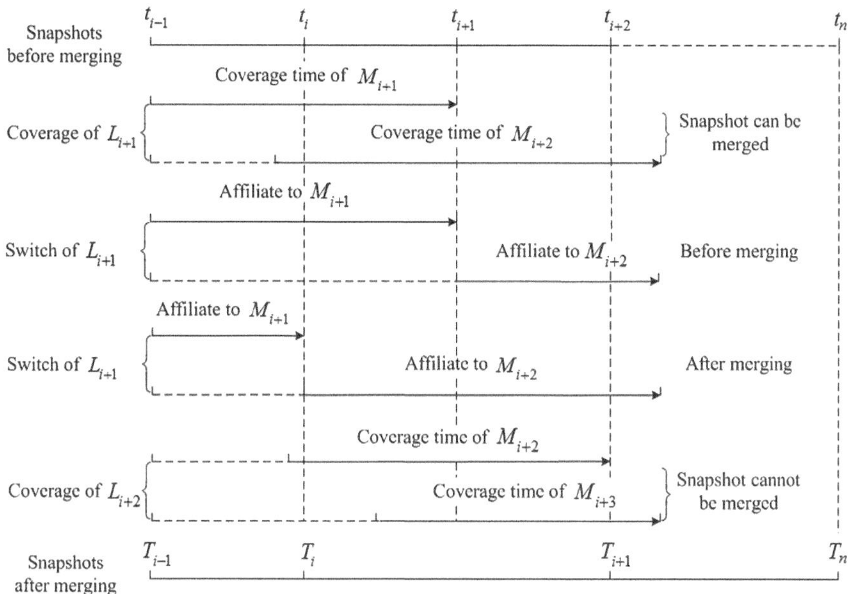

Fig. 3.8 Principal of snapshot merging

Table 3.1 Snapshot merging algorithm

Algorithm 3.1 Snapshot merging algorithm in the system cycle

```
Let  T={ t1, …,ti, …, tn } be the switching times in SGRP;
     L={ L1, …,Li, …, Ln } be the LEO satellite which switches at time ti;
     Fi=[xi, yi] be the free period of satellite Li;
     𝔍={T1, …,Ti, …, Tm }be the time division of snapshot after merging;
     k=1, T1= t1;
for (i=1;  i≤n;  i++)
{
     if (Tk ∈Fi+1 )
        continue;
     else
     {
         k++;
         Tk = ti+1;
      }
 }
```

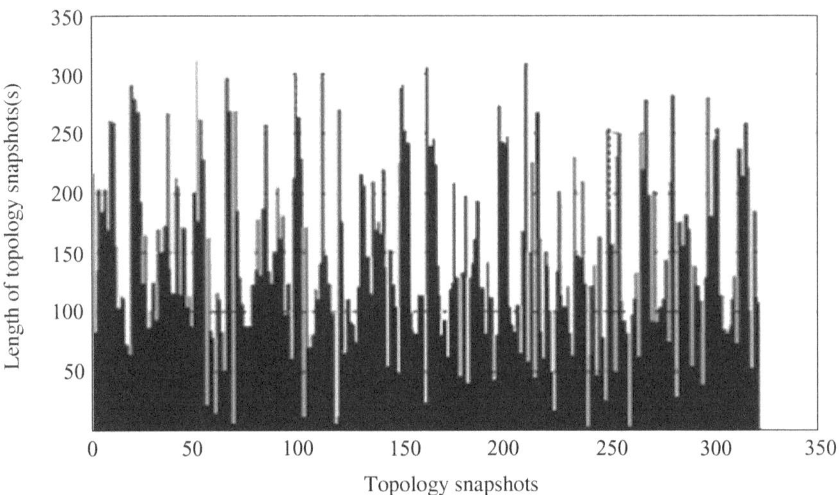

Fig. 3.9 Topology snapshots after merging

cannot converge in too short snapshots. If there is are a large number of snapshots in the system cycle, the stability of the entire system will be poor, and large memory is required to store all these snapshots. The improvement of snapshot division eliminates the snapshot debris, and greatly reduces the number of snapshots in the system cycle. This provides sufficient time for the QoS routing algorithms to converge, and hence increases the stability of the system.

3.3.2.3 Generation of LMR

In the beginning of each snapshot, the LEO satellite $L_{i,j,k}$ needs to report its layer management $LMR(L_{i,j,k})$ to the MEO satellite $GM(L_{i,j,k})$ for which to calculate the routing table. The MEO satellite $M_{i,k}$ also needs to report its $LMR(M_{i,k})$ to the GEO satellite $GM(M_{i,k})$ for which to monitor and manage the entire system. The *LMR*s of LEO and MEO satellites are calculated as follows:

(1). *LMR* of the LEO satellite: The LEO satellite $L_{i,j,k}$ connects to ground station $T^t_{L_{i,j,k}}\left(0 \le t < S_T\left(L_{i,j,k}\right)\right)$ through $UDL_{L_{i,j,k} \to T^t_{L_{i,j,k}}}$. The ground station $S_T\left(L_{i,j,k}\right)$ returns the number of ground stations that are connected to $L_{i,j,k}$. $L_{i,j,k}$ also connects with four adjacent LEO satellites (The LEO satellites on both sides of the reverse seam connect with only three adjacent LEO satellites), marked as $\left\{L_{i,j,k0}, \ldots, L_{i,j,k3}\right\}$. Besides, the LEO satellites can also connect with the MEO and GEO satellites through IOL. The LMR of an LEO satellite can be obtained as follows:

$$LMR(L_{i,j,k}) = \left\{(A,\ C(UDL_{L_{i,j,k}\to A})|A = T^{0}_{L_{i,j,k}}, \ldots, T^{S_T(L_{i,j,k})-1}_{L_{i,j,k}}\right\}$$
$$\cup\left\{(B,\ C(ISL_{L_{i,j,k}\to B})|B = L_{i,j,k0}, \ldots L_{i,j,k3}\right\}$$
$$\cup\left\{(C, C(IOL_{L_{i,j,k}\to C})|C = M_{i,j}, G_i\right\}$$

(2). *LMR* of the MEO satellite: The MEO satellite $M_{i,j}$ connects to the ground base station $T^{t}_{M_{i,j}}(0 \le t < S_T(M_{i,j}))$, and also the adjacent MEO satellites $\{M_{i,j_0}, \ldots, M_{i,j_n}\}$, in which n is the number of adjacent MEO satellites. $S_G(M_{i,j})$ returns the number of elements in $GL(M_{i,j})$. $M_{i,j}$ communicates with GEO and LEO satellites through IOL. The *LMR* of MEO satellite can be obtained as follows:

$$LMR(M_{i,j}) = \left\{(A,\ C(UDL_{M_{i,j}\to A})|A = T^{0}_{M_{i,j}}, \ldots, T^{S_T(M_{i,j})-1}_{M_{i,j}}\right\}$$
$$\cup\left\{(B,\ C(ISL_{M_{i,j}\to B})|B = M_{is,js},\ s = 0, \ldots, n-1\right\}$$
$$\cup\left\{(C,\ C(IOL_{M_{i,j}\to C})|C = L_{i,j,k},\ k = 0, \ldots, S_G(M_{i,j}) - 1\right\}$$
$$\cup\left\{(G_i,\ C(IOL_{M_{i,j}\to G_i})\right\}$$

3.3.2.4 Spreading of LMR

The LMRs of LEO and MEO satellites need to be spread throughout the entire satellite network system after generation. According to the description of LMR above, the spreading steps of LMR are as follows:

Step 1 The LEO satellite $L_{i,j,k}$ measures the states of its outgoing links in the beginning of each snapshot, and generates *LMR*($L_{i,j,k}$) according to the definition.

Step 2 *LMR*($L_{i,j,k}$) is reported to *GM*($L_{i,j,k}$) through IOL. The MEO satellite waits for a period of time δ. If the number of *LMRs* *GM*($L_{i,j,k}$) received during the period δ is equal to the number of elements in $GL(GM(L_{i,j,k}))$, go to step 3; otherwise, *GM*($L_{i,j,k}$) will send satellite failure report *alert*(*Y*) to its manger GEO satellite $GM(GM(L_{i,j,k}))$, and then go to step 3.

Step 3 The MEO satellite $M_{i,j}$ measures the states of its outgoing links in the beginning of each snapshot, and generates *LMR*($M_{i,j}$) according to the definition.

Step 4 When an MEO satellite receives all the *LMR*($L_{i,j,k}$), it will send them together with its *LMR*($M_{i,j}$) to its adjacent MEO satellite in the same orbit plane. When an MEO satellite moves to the vicinity of the ascending/descending cross point of the two MEO orbit planes, it will transmit all the *LMR*($L_{i,j,k}$) and *LMR*($M_{i,j}$) to the adjacent MEO satellite in the adjacent MEO orbit through inter-orbit ISL. As a result, each MEO satellite can master the link and topology status of the whole satellite network, marked as LMR_{whole}.

Step 5 The MEO satellite $M_{i,j}$ sends LMR_{whole} to the GEO satellite $GM(M_{i,j})$. The GEO satellite will broadcast the LMR_{whole} in the GEO layer and forward it to the ground stations in its coverage area.

3.3.2.5 Calculation of Ordinary Routing Table

In the NSGRP, the routing service can be classified into ordinary routing and QoS routing. Ordinary routing uses delay as the optimization objective. It is provided by the MEO satellites. QoS routing will be discussed in the subsequent chapters.

In order to make the routing structure of a multilayered satellite network clear, not all the link status information in *LMR* is used for routing computation. The calculation of ordinary routing table is based on the following assumptions:

(1) During data transmissions, the data packets from the ground gateway enter the satellite network system only through the LEO interface satellite. The ground gateway does not communicate with MEO or GEO satellites directly.
(2) During data transmissions, the data packets from the satellite network leave the network only through the LEO interface satellite. The MEO or GEO satellites don't communicate with the ground gateway directly.
(3) During data transmissions, the LEO satellites only connect with MEO satellites and the ground gateway, and the GEO satellites are not involved in data forwarding when the traffic load is normal.
(4) The MEO satellites not only need to calculate the routing table for the LEO satellites, but also need to forward data for the LEO satellites.
(5) The MEO satellite M_{ij} is only in charge of the routing computation for the LEO satellites in $GL(M_{i,j})$.

The ordinary routing table computation is carried out on in three steps:

Step 1 When an MEO satellite M_{ij} receives all the *LMR*s of the LEO satellites in $GL(M_{ij})$ and the *LMR*s in the same MEO layer, it will use the end-to-end delay as the optimization objective and adopt the SPF algorithm to calculate the paths $P_{X\to Y}$ from the LEO satellites in $GL(M_{ij})$ to all the other LEO satellites. The results will be added to $LOT(M_{i,j}|X)$

Step 2 The MEO satellite M_{ij} uses the end-to-end delay as the optimization objective and adopts the SPF algorithm to calculate the paths from itself to all the other LEO/MEO satellites. The results are to added to $MRT(M_{i,j})$

Step 3 Put the second column and the last column of $LOT\left(M_{i,j}|X\right)$ into $LRT(X)$, and send it to LEO satellite X through $\text{IOL}_{M_{i,j}\to X}$.

3.3.2.6 Congestion Avoidance

Compared with SGRP, the NSGRP improves the congestion avoidance strategy. In the NSGRP, LU is used to judge whether a link is congested. Since the propagation delay of the ISLs in MEO layer is relatively large and less frequently used in routing service, it usually has less congestion. When an MEO satellite $M_{i,k}$ in MEO group M_i finds the LU of its outgoing links ≥95 %, it will send *alert*() to the GEO satellite $GM(M_{i,k}) = G_i$. The GEO satellite will then bypass the congested links. When an LEO satellite $L_{i,j,k}$ finds the LU of its outgoing links ≥95 %, it will send *alert*() to the MEO satellite $GM(L_{i,j,k}) = M_{i,j}$. The mechanism of LEO layer congestion avoidance then starts as follows:

Step 1 Assume that link $\mathrm{ISL}_{L_{i,j,p} \to L_{i,j,q}}$ is congested. When the MEO satellite $GM(L_{i,j,p})$ receives the $alert\left(\mathrm{ISL}_{L_{i,j,p} \to L_{i,j,q}}\right)$, it first calculates $CGA\left(\mathrm{ISL}_{L_{i,j,p} \to L_{i,j,q}}\right)$, and then informs the MEO layer about the result. The MEO satellite receiving the $CGA\left(\mathrm{ISL}_{L_{i,j,p} \to L_{i,j,q}}\right)$ will set the delay of $\mathrm{ISL}_{L_{i,j,p} \to L_{i,j,q}}$ to ∞

Step 2 In order to save the limited onboard computing resources of the MEO satellites, NSGRP only reroutes the paths that are affected by $CGA\left(F_{L_{i,j,p} \to L_{i,j,q}}\right)$. The pseudo code of the reroute algorithm is shown in Table 3.2.

In many cases, link congestions are temporary. After rerouting, the LUs of links in *CGA*() will decrease. If the routes are not reset accordingly, the links in *CGA*() will be idle, which results in a waste of resources. When an LU of link in *CGA*() decreases below a certain threshold, the original routing table can be recovered by the path fragment stored in *temp*. This is also an improvement compared with SGRP.

3.3.2.7 The Process of Node Failure

The satellites in a satellite network may fail or cease working for many reasons, such as maintenance or testing, temporary or permanent hardware failure, antenna closure when crossing the Polar Regions or oceans to save energy. When a satellite fails, to ensure that the data packets in the paths that pass through the satellite are delivered reliably, all these paths should be rerouted. In NSGRP, the treatments are different when a failure happens in different layers.

1. LEO satellite failure: When an LEO satellite $L_{i,j,k}$ fails, its neighbor $L_{i,j,k+1}$ (usually is the nearest LEO satellite or the adjacent satellite in the same orbit plane) that first monitors the occurrence of the failure, will send the failure report $alert\left(L_{i,j,k}\right)$ to the MEO satellite $GM\left(L_{i,j,k+1}\right) = M_{i,j}$. After receiving the

Table 3.2 Rerouting algorithm of LEO layer

Algorithm 3.2 Rerouting algorithm of LEO layer

Assume There are m rows in matrix $LOT(M_{i,j}|X)$, the third element of triples <X, Y, $P_{X \to Y}$> in the ith row is the set of links in column n_i, which is marked as (x, y);

// The matrix *temp* has m rows, and is used to store the path fragment before rerouting for recovery purposes.

```
Set all the elements in CGA(ISL_{Li,j,p→Li,j,q}) ∞;
for (i=1;  i≤m;  i++)
{
    k=0; // Starting point flag of rerouting
    for (j=1;  j≤m;  j++)
    {
        if (LOT(i,j)∈ CGA(ISL_{Li,j,p→Li,j,q}))
        {
            if (k==0)
            {
                k=1;
                h=j; //Record the starting point of rerouting
                j→temp(i ); //Store the starting point of rerouting
                continue;
            }
        }
        else
        {
            if (k==0)
              continue;
            else
            {
                h→temp(i ); // Store the starting point of rerouting
                path(h,j)→temp(i ); //Store original path fragement
                reroute(fe(LOT(i,h)), fe(LOT(i,j)));
                k=0;
            }

        }
    }

}
/* path(h,j) returns the path from LOT(i,h) to LOT(i,j);
fe() returns the first element of the tuple;
 reroute() is the SPF rerouting algorithm; */
```

report, $M_{i,j}$ sets the weights of all $\mathrm{ISL}_{X \to L_{i,j,k}}$ and $\mathrm{ISL}_{L_{i,j,k} \to X} \infty$, where $X \in GL(M_{i,j})$, and informs the GEO satellite $GM(M_{i,j}) = G_i$ and all the other MEO satellites to reroute. The failure report $alert(L_{i,j,k})$ is broadcasted to the ground base station all over the world through the GEO satellites. During the period of an LEO failure, its group manager MEO satellite fills its position.

2. MEO satellite failure: When an MEO satellite $M_{i,j}$ fails, the nearest MEO satellite $M_{i,j+1}$, which first monitors the occurrence of the failure, will send the failure report $alert(M_{i,j})$ to the GEO satellite $GM(M_{i,j+1}) = G_i$. The $alert(M_{i,j})$ is broadcasted by G_i in the GEO layer, and all the MEO satellites and LEO satellite in $L_{i,j}$ are informed that the MEO satellite $M_{i,j}$ is replaced by the GEO satellite G_i temporarily. After that, all the ground base stations will be informed that the MEO satellite $M_{i,j}$ has failed. Since the propagation delay of the IOL between the LEO layer and the GEO layer is large, the system performance will deteriorate during the period of the MEO satellite failure. Besides, an MEO layer satellite plays an important role in NSGRP, for it is in charge of routing calculations, group management and data forwarding. The MEO satellite should be repaired or replaced as soon as possible after the failure, so as not to affect the performance of the entire system.
3. GEO satellite failure: When a GEO satellite G_i fails, the nearest GEO satellite G_{i+1}, that first monitors the failure, will send an $alert(G_i)$ through the links of the MEO layer to the MEO satellites in group M_i. The MEO satellites that have received the report will not report to G_i. The failure report is broadcasted in the GEO layer, and then forwarded to all the ground stations. The GEO satellites play less important role than the LEO and MEO satellites in NSGRP. Its main function is to monitor the network and acts as a backup of the MEO and LEO satellites. However, the GEO satellites may contribute to an early detection of an MEO and LEO satellite failure, and broadcast the failure to all the ground stations. Therefore, the GEO satellite failure should also be repaired as soon as possible so as not to affect the robustness and survivability of the satellite network system.

The SGRP strategy does not work even when a single MEO satellite fails. However, the NSGRP strategy can work well under such circumstances since a GEO satellite can take the place of a failing MEO satellite. This greatly increases the robustness of the whole system. The NSGRP strategy does not work only when an MEO satellite and two GEO satellites that both cover it simultaneously fail to work, which actually seldom happens.

3.4 The Structure of NSGRP

Having discussed the basic concepts, the basic components and the implementation of the NSGRP strategy in the above Sects. 3.1–3.3, this section will focus on the structure of NSGRP, so that readers can fully understand NSGRP.

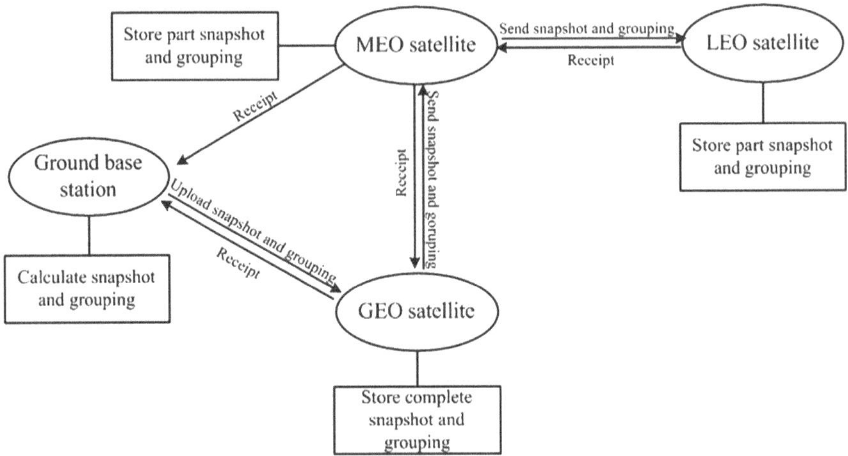

Fig. 3.10 The function module structure of the establishing process of NSGRP

The implementation of NSGRP consists of four processes: the establishing process, the initializing process, the data forwarding process, and the exception handling process. The processes will be described in detail as follows.

1. Establishing process

The establishing process is the work NSGRP needs to do during the initial stage of the *Tr* constellation, when all the satellites in the constellation are launched into the preset orbits and all the links are established. The function module structure of NSGRP in the establishing process is shown in Fig. 3.10, in which a rectangle represents a function module and an ellipse represents a physical component. Since the memory of the LEO and MEO satellites is limited, the GEO satellite may send only a part of snapshots and grouping to the MEO satellite according to its size of memory. Similarly, the MEO satellite may send only a part of snapshots and grouping to the LEO satellite. The snapshots division and grouping keep unchanged if there is no satellite failure.

2. Initializing process

The initializing process is the work each physical component of NSGRP needs to do at the beginning of each snapshot. The function module structure of NSGRP in the initializing process is shown in Fig. 3.11.

The *LMR* in Fig. 3.11 stored in an MEO satellite is used for routing calculation, and can be discarded after the routing computation is done. The *LMR* stored in a GEO satellite is used to compute the *MRT* of an MEO satellite when it fails. The GEO satellite will replace the MEO satellite immediately in this way. The LEO satellite collects all the *LMR*s that are related to itself and then reports to its MEO manager satellite. The *LMR*s are not stored in the LEO satellite. When the LEO satellite receives the *LRT* sent by its MEO manager satellite, it informs the ground-based station that data transmission is approved.

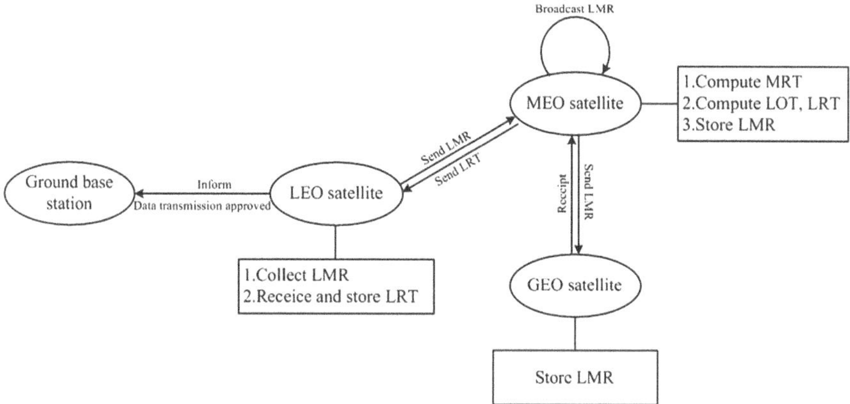

Fig. 3.11 The function module structure of the initializing process of NSGRP

The initializing process of NSGRP may restart with the failure of satellite nodes. However, the failure of a GEO satellite does not affect the division of the snapshots, and the initializing process doesn't need to restart during repairs. When an LEO satellite fails, the ground base station will recalculate the snapshot division before the satellite is repaired. When an MEO satellite fails, if its manager GEO satellite is able to replace this MEO satellite, then the snapshot division doesn't need to be recalculated, and the initialization process will not be affected; otherwise, the snapshots need to be recomputed, the initialization process hence needs to restart according to the new snapshot division.

The MEO satellite can send *LRT* to the LEO satellites, and at the same time, send *MRT* and *LOT* to the GEO satellites. This eliminates the overhead of recalculation of GEO satellites, but increases the communication overhead and transmission of MEO satellites. In the implementation of NSGRP, the two schemes can be selected according to the type and ability of MEO and GEO satellites.

3. Data forwarding process

The data forwarding process is the process that the LEO satellites accept the data transmission requirements and forward the packets according to *LRT* when they are ready. The function module structure of NSGRP in the data forwarding process is shown in Fig. 3.12.

In Fig. 3.12, the ground base station sends the data packets to the LEO satellite bound to its logical location, which is also called the interface satellite. The interface satellite will forward the packets to the next hop satellite according to the destination address and *LRT*. The process is repeated until an LEO satellite finds that its logical address is the same as the destination logical address. This satellite is the exit satellite and will forward the packets to the ground base station in its coverage area according to the destination address.

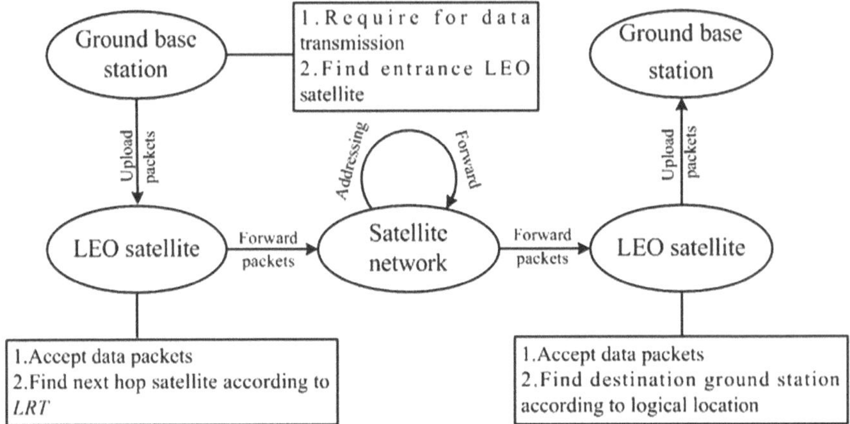

Fig. 3.12 The function module structure of the data forwarding process of NSGRP

4. Exception handling process

The exception handling process includes work each physical component needs to do when a congestion or failure occurs mentioned in Sect. 3.3.2. As mentioned before, the propagation delays of the IOL between the LEO and MEO layers and the ISL within the MEO layer are relatively large, so IOL and ISL are selected in low probability in the routing computation, hence are idle most of the time. As a result, only the congestion of the LEO layer ISL is considered. The handling process of $IOL_{LEO \to MEO}$ and MEO layer ISL is similar to that of the LEO layer, hence no tautology is needed here.

In the satellite network system, the possibility of a node failure is less than the possibility of link congestion, and the node failure possibility of the LEO and MEO satellites with a heavier burden is larger than the GEO satellites with a lighter burden. The details of node failure handling were already discussed in Sect. 3.3.2. Some function module structures of the handling process are given below. First, let's discuss the function module structure of the link congestion handling process, as shown in Fig. 3.13.

The report *resume*() in Fig. 3.13 is used to inform the MEO satellites that congestion is relieved, hence the former *MRT*, *LOT* and *LRT* can be recovered. Since the former routing table is stored in *temp*() (as mentioned in Algorithm 3.2), the NSGRP can restore the previous routing table quickly.

The function module structure of the node failure handling process is shown in Fig. 3.14. In Fig. 3.14, condition 2 of the MEO function module unconditionally transfers to condition 3 of the MEO function module. Similarly, condition 2 of the GEO function module unconditionally transfers to condition 3 of the GEO function module. This means that the MEO satellite recalculates the routing immediately when it receives a failure notice, while the GEO satellite broadcasts to the ground stations when received the failure notice.

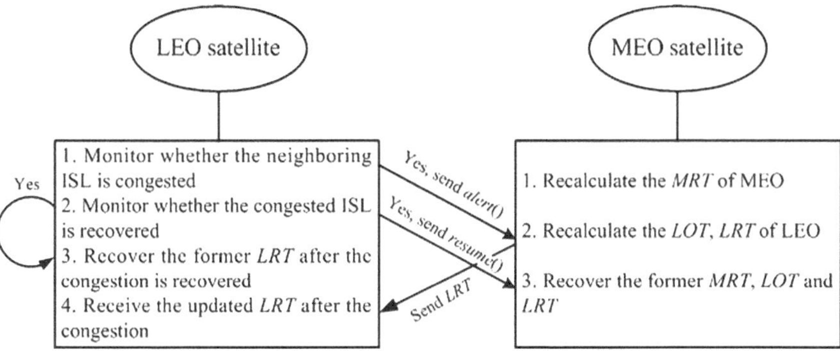

Fig. 3.13 The function module structure of the congestion handling process of NSGRP

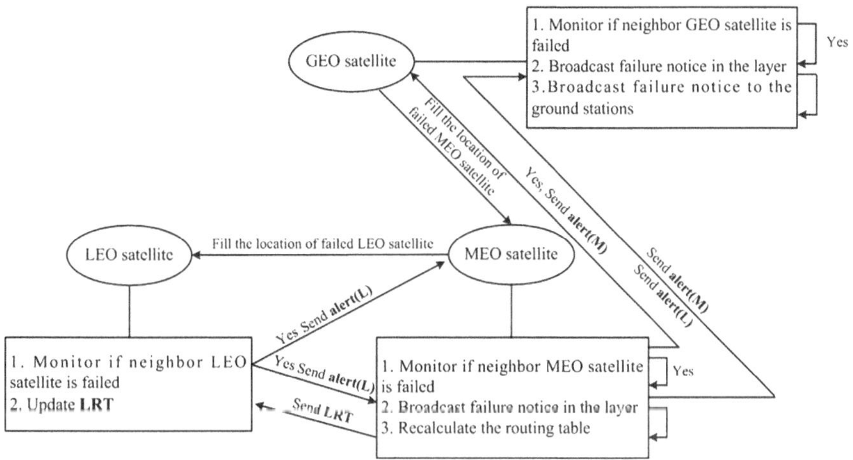

Fig. 3.14 The function module structure of the node failure handling process of NSGRP

3.5 Simulations and Results

Since the main comparison object of NSGRP is SGRP, in order to make the performance comparison fair and meaningful, this text adopts similar parameters and simulation environments as described in [17]. The NS and STK are selected as our simulation tools. Same as the SGRP experimental parameters, here the bandwidth of all UDLs and ISLs is set at 160 Mbps, and the cache of an outgoing link is 5 MB. Assuming that the average packet length is 1000 byte, it is easy to know that the link bandwidth is 20000 packets/s, and the cache size is 5,000 packets. The delay sampling time is 1 min. First, we model the *Tr* constellation with STK, and its backbone LEO/MEO layer is shown in Fig. 3.15.

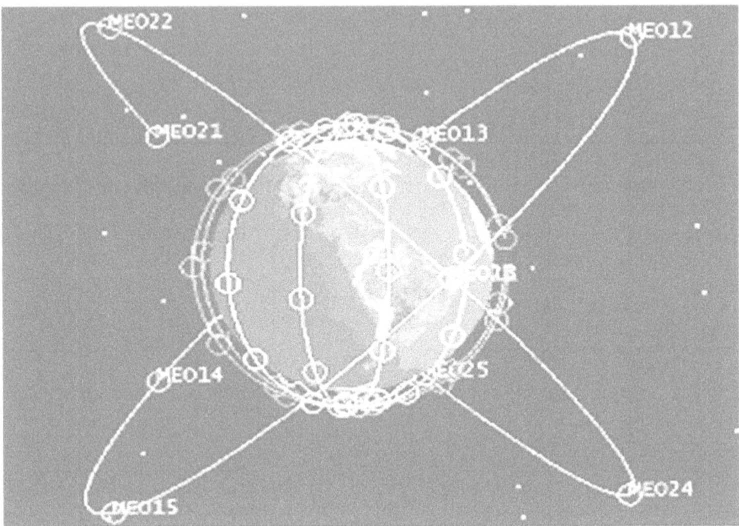

Fig. 3.15 STK model of LEO/MEO layer of *Tr* constellation

In the simulation, a total of three kinds of protocols involve in the comparison: NSGRP, SGRP, and the optimal routing protocol. The optimal routing protocol uses Dijkstra algorithm to calculate routes, and assumes that each satellite knows the topology of the whole network, and the link status is updated in real-time. It is an ideal state with no protocol overhead, and is the performance limit of all routing protocols.

Our performance comparison is based on the end-to-end delay of several OD ground stations. The difference between SGRP and NSGRP is that an LEO/MEO double-layered satellite constellation is used in SGRP and the *Tr* constellation and the optimal routing protocol is used in NSGRP. To compare the performance of these three protocols in different situations, three experimental scenes are designed.

1. Delay characteristic: The first set of simulations shows the end-to-end delay characteristic of the three routing protocols.
2. Effects of node failure: The second set of simulations shows the performance of the three routing protocols when a node failure occurs.
3. Effects of link congestion: The third set of simulations shows the performance of the three routing protocols when a link is congested.

3.5.1 Delay Characteristic

This part compares the delay characteristic of these three protocols. To be consistent with the parameters in SGRP, we choose three pairs of OD ground stations. The first two pairs have the same source ground station that is located in Asia (N:

37.5°, E: 112.5°). Their destination ground stations are located in North America (N: 33.25°, W: 277.5°) and Europe (N: 52.5°, E: 52.5°), separately. The space base path between these two stations crosses a busy business region [17]. The third pair of ground stations are located in Oceania (S: 37.5°, E: 142.5°) and Africa (S: 18.25°, E: 37.5°) separately, where there is far less business. To each pair of OD ground stations, the source station sends a continuous flow of information at the average flow rate of 8 Mb/s during a period of 100 min.

To compare the performances of the routing protocols under different loads, we gradually increase the LUs of the ISLs in the LEO layer. First, we generate a flow with an exponentially distributed data rate whose average value is λ. This flow is mapped into the ISLs in the LEO layer according to the shortest path first routing. And then, the value λ is gradually increased to increase the LUs of the ISLs. It is assumed here that all the satellite link queuing delays meet the M/M/1 model.

The simulation time is 100 min. The delay characteristics of the optimal routing, SGRP and NSGRP are shown in Fig. 3.16. For each average link load, the end-to-end delay is the average delay during the 100 min. Since the space-based paths of some OD ground stations cross busy business regions, link congestion is observed even when the link load is 5 %.

As shown in Fig. 3.16, the performances of the three protocols are nearly the same even in those paths that cross the regions with high traffic density (Fig. 3.16a, b) when the link load is less than 5 %. This is because the propagation delay dominates the end-to-end delay when the link load is not high. However, there are obvious differences between NSGRP, SGRP and the optimal routing protocol when the average load is about 8 %. The reason is that as the ISL load increases, the ISLs with higher traffic density are more prone to get congested, and the queuing delay and processing delay dominate the end-to-end delay in a congested environment.

For the paths not crossing the traffic intensive area, as shown in Fig. 3.16, there is little difference between SGRP, NSGRP and optimal routing in the average end-to-end delay when the link load is less than 50 %. According to the experiment report, the link congestion in the third experiment is lighter than the first two experiments. Hence the difference between SGRP and NSGRP is less obvious than the first two experiments, about within 0.3 ms.

NSGRP performs well in all the three experiments in Fig. 3.16. The reasons may be as follows: First, the number of snapshots in SGRP is too large, which makes frequent link switching and large protocol overhead. This is not conductive to the implementation of routing algorithms; second, NSGRP is based on the *Tr* constellation, in which the MEO and GEO layers can carry part of the traffic when congestion occurs; third, there is no congestion recovery mechanism in SGRP. The recovered links can not be used in the current snapshot, which reduces the available ISLs and even causes new congestion.

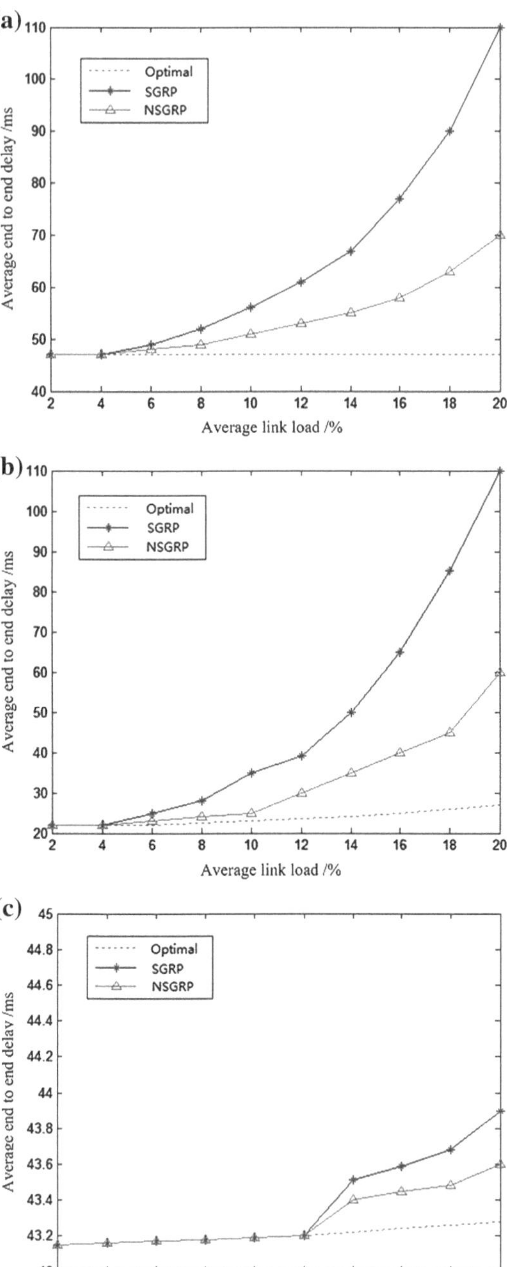

Fig. 3.16 Comparison of average end-to-end delay **a** The first OD pair of ground stations; **b** The second OD pair of ground stations; **c** The third OD pair of ground stations

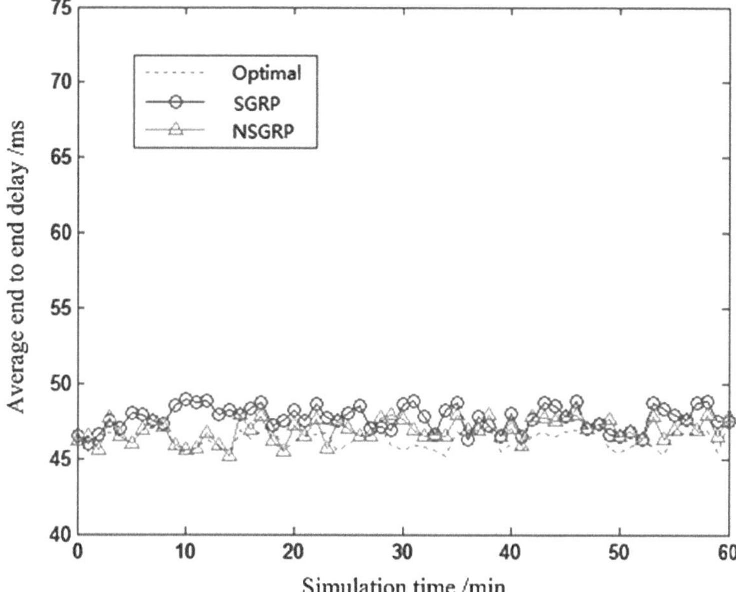

Fig. 3.17 Comparison of average end-to-end delays during node failure

3.5.2 Effects of Node Failure

There are node failure handling strategies in both SGRP and NSGRP. In the following paragraphs, we will compare the end-to-end delays of these two protocols when there is a node failure. To reflect the real-time changes of the delay, and to be consistent with the background traffic of SGRP, we modify our background traffic according to the traffic characteristic during different times of a day. The detailed modification is shown in [17].

The node failure may affect the routing and path delay. In this simulation, the end-to-end delays of the first OD pair of ground stations using the optimal, SGRP and NSGRP protocols are tracked. The source ground station produces 1000 packets/s for 60 min. The simulation period is set from 8:00 to 9:00 am local time. The average end-to-end delays of the three protocols during a node failure are shown in Fig. 3.17.

As shown in Fig. 3.17, the performances of SGRP and NSGRP are similar when a satellite node fails. The disadvantage of DRA that a satellite failure cannot be discovered in time is eliminated in both SGRP and NSGRP. Hence the effect of satellite failures on SGRP and NSGRP is little. The failure report is delivered to the GEO and MEO layers in time so that the whole network will know the failure immediately. The overall performance of SGRP is slightly less favorable than that of NSGRP, for the satellite in the higher layer will replace the failed satellite in the lower layer in NSGRP while such mechanism is absent in SGRP.

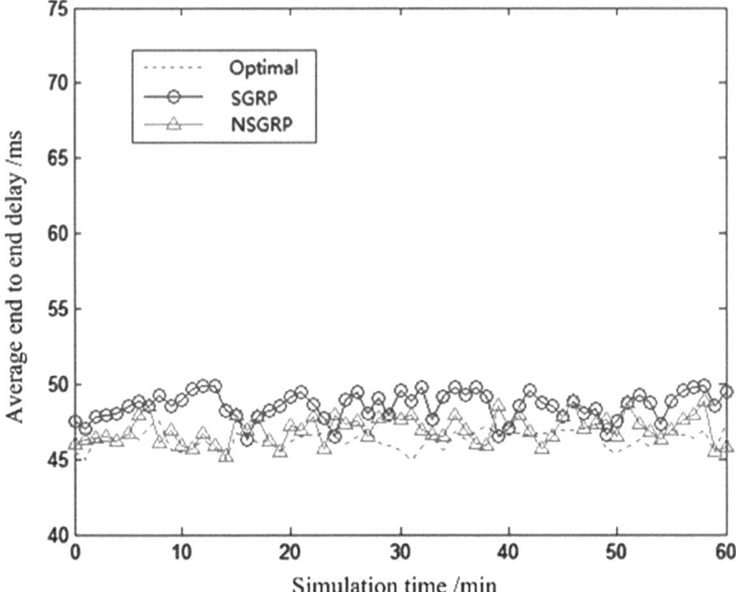

Fig. 3.18 Comparison of average end-to-end delays when link is congested

3.5.3 *Effects of Link Congestion*

For a comparison of the link congestion handling protocols, we also choose to compare the average end-to-end delays between the first OD pair of ground base stations using optimal, SGRP and NSGRP. The link congestion is caused by heavy traffic in certain regions in the satellite network. In this simulation, the source station located (N: 37.5°, E: 112.5°) in Asia generates 1000 packets/s in a duration of 60 min during the peak traffic period 10:00–11:00 am (local time). And then the congested links with the LU equal to 100 % are observed. The average end-to-end delays of these three protocols during a link congestion are shown in Fig. 3.18.

The simulation results show that the average difference between the SGRP and the optimal is 0.75 ms, while the difference between the NSGRP and the optimal is 0.5 ms. The routing tables are recalculated immediately in both NSGRP and SGRP. The lack of the congestion recovery mechanism may be the reason why SGRP performs slightly unsatisfactorily compared to NSGRP when a congestion occurs.

3.6 Summary

A new satellite network routing protocol-NSGRP is proposed in this chapter based on the SGRP. The NSGRP improves the snapshot division of SGRP, and greatly reduces the number of snapshots in a system cycle, which is conducive to the implementation of routing algorithms. At the same time, the NSGRP improves the link congestion and node failure handling mechanisms, making the routing strategy more robust. Lastly, the performances of SGRP and NSGRP are compared using the same experimental parameters and environment used by SGRP. The simulation results show that, NSGRP performs better than SGRP no matter whether the traffic is normal, or when a link is congested or a node fails.

References

1. Werner W, Jahn A, Lutz E et al (1995) Analysis of system parameters for LEO/ICO satellite communication networks. IEEE J Sel Areas Commun 13(2):371–379
2. Wood L, Clerget A, Andrikopoulos I et al (2001) IP routing issues in satellite constellation networks. Int J Satell Commun 19(2):69–92
3. Ekici E, Akyildiz IF, Bender MD (2001) A distributed routing algorithm for datagram traffic in LEO satellite networks. IEEE/ACM Trans on Netw 2:137–147
4. Berndl G, Werner M, Edmaier B (1997) Performance of optimized routing in LEO intersatellite link networks. In: Proceedings of IEEE 47th vehicular technology conference (VTC'97), vol 5. Phoenix, pp 246–250
5. Uzunalioglu H, Akyildiz IF, Bender MD (2000) A routing algorithm for connection-oriented low earth orbit (LEO) satellite networks with dynamic connectivity. Wirel Netw 6(3):181–190
6. Manger R, Rosenberg C (1997) A QoS guarantees for multimedia services on a TDMA-based satellite network. IEEE Commun Mag 35(7):56–65
7. Uzunalioglu H, Akyildiz IF, Yesha Y et al (1999) Footprint handover rerouting protocol for low earth orbit satellite networks. Wirel Netw 5(5):327–337
8. Tsai K, Ma R (1995) Darting: a cost effective routing alternative for large space-based dynamic topology networks. In: Proceedings of IEEE military communications conference (MILCOM'95), vol 10. McLean, pp 682–687
9. Ekici E, Akyildiz IF, Micheal DB (2000) Datagram routing algorithm for LEO satellite networks. In: Proceedings of IEEE INFOCOM'2000, Tel-Aviv, pp 500–508
10. Werner M (1997) A dynamic routing concept for ATM-based satellite personal communication networks. IEEE J Sel Areas Commun 8:1636–1648
11. Chang HS, Kim BW, Lee CG et al (1998) FSA based link assignment and routing in Low Earth Orbit satellite networks. IEEE Trans Veh Technol 8:1037–1048
12. Lee J, Kang S (2000) Satellite over Satellite (SOS) network: a novel architecture for satellite network. In: Proceedings of IEEE INFOCOM'2000, vol 3, issue 1. Tel-Aviv, pp 315–321
13. Hu JH, Yeung KL (2000) Routing and rerouting in a LEO/MEO two-tier mobile satellite communications system with inter-satellite links. In: Proceedings of IEEE international conference on communications (ICC'00) vol 1. New Orleans, pp 134–138
14. Akyildiz IF, Ekici E, Bender MD (2002) MLSR: a novel routing algorithm for multi-layered satellite IP networks. IEEE/ACM Trans Netw 3:411–424

15. Chen C, Ekici E, Akyildiz IF (2002) Satellite grouping and routing protocol for LEO/MEO satellite IP networks. In: Proceedings of the 5th ACM international workshop on wireless mobile multimedia (WoWMoM'2002) vol 9. Rome, pp 109–116
16. Chen C, Ekici E (2005) A routing protocol for hierarchical LEO/MEO satellite IP networks. ACM Wirel Netw J (WINET) 11(4):507–521
17. Chen C (2005) Advanced routing protocol for satellite and space networks. Ph.D. thesis, Georgia Institute of Technology, Atlanta

Chapter 4
Satellite Network Traffic Engineering

4.1 Introduction

Traffic engineering is a method to optimize the performance of a satellite network through dynamically analyzing, forecasting, and balancing the network traffic [1]. It mainly deals with the problem of how to allocate the network traffic, and determines the efficiency, robustness, and reliability of the network [19]. Traffic engineering has been widely used in terrestrial networks. However, there have been relatively few discussions about satellite network traffic engineering. Actually, because of their limited resources and inconvenience in maintenance, satellite networks are more in need of traffic engineering than terrestrial networks to improve the operational efficiency of the network and to extend the life span of the system.

There are a lot of discussions on traffic engineering in terrestrial networks, among which the essays [3–12] are most representative. The traffic engineering methods in these essays use traffic demand matrices and network topologies as the input, and optimize the performance of the network through properly allocating the traffic. There are two major challenges in traffic engineering in a dynamic environment: uncertainty of the traffic and uncertainty of the network topology.

The uncertainty of the traffic refers to some unexpected traffic in the network. For many autonomous systems (AS) of the terrestrial network, although the traffic is relatively stable most of the time, there may be some periods of great changes in traffic, including some periods with unpredictable traffic spikes. Hao found that network traffic can increase by one order of magnitude in a short period of time after tracking the traffic of several major U.S. backbone networks. Later, network engineers confirmed that this burst of traffic was not an accidental isolated incident after investigating the records of previous years [13].

The uncertainty of the network topology means topology changes or network failures caused by accidents, maintenance errors, natural disasters, or even malicious attacks [14, 15]. Such as a cable cut-off caused by digging mistakes, the most common physical layer accident, happens almost every a few months. The router

F. Long, *Satellite Network Robust QoS-aware Routing*,

DOI: 10.1007/978-3-642-54353-1_4,

failure caused by configuration mistakes happens even more frequently. The uncertainty of the network topology increases difficulty in traffic engineering.

For uncertain traffic, most current traffic engineering algorithms belong to the prediction-based traffic engineering method, which does not cope with sudden traffic. The usual practice of this kind of method is to collect a sample set of traffic and then optimize the traffic allocation based on this sample set. The algorithm may optimize for the average or the worst case of this set. The advantage of this kind of algorithm is that the algorithm can achieve near optimum performance when the network traffic is basically stable and the actual traffic is similar to the sample traffic. However, the performance of the algorithm will deteriorate sharply when the unpredicted traffic spike deviates a lot from the sample set, as the algorithm only optimize for the traffic sample set. An extreme case of prediction-based traffic engineering is online adaptation. The advantage of this kind of algorithm is that if it can converge quickly, it only needs to collect a small sample set. However, such a kind of algorithm can experience a large transient penalty when there are fast and significant traffic changes.

Hao et al. observed the real Abilene topology and its traffic traces and proved that for bottleneck links, the traffic intensity generated by all the three prediction-based algorithms that they evaluated exceeds the link capacity during traffic spikes, and some reaches 2.44 times the link capacity, while for an optimal algorithm, no link receives traffic above 50 % of its capacity [13]. One of the means to cope with the unpredictable traffic spike is the oblivious routing method as described in [3, 4, 10, 12, 16]. In oblivious routing, the algorithm is independent of the sample set traffic matrix, hence has the potential to handle traffic spikes well. The main idea of oblivious routing is to assign a flow split parameter α_j to each node which is independent of the destination of the flow. Each node allocates the traffic according to the flow split parameter α_j. The selection of the parameter usually meets certain conditions, and the traffic spike is considered. In this way, a fixed bandwidth channel is formed among the nodes, which makes routing independent of the traffic matrix. The essence of oblivious routing is to optimize for the traffic spike together with normal traffic. So it performs lower than the prediction-based algorithms when the traffic is normal.

The topology uncertainty of the satellite network does not need to be considered. This is because the number of satellite routers in the satellite network is much smaller than the terrestrial routers. The topology variability of the satellite network itself has been shielded by NSGRP already. The topology uncertainty of the terrestrial network refers to router failure caused by fault, attacks, or disaster. For the satellite network, a small amount of onboard router failure can be handled by the node failure handling mechanism as described in Sect. 3.3.2. When there are too many failed onboard routers, causing the system to operate abnormally, the traffic engineering of the terrestrial network cannot be used in the satellite network either. Therefore, we only consider the uncertainty of the traffic in traffic engineering of the satellite network.

For satellite network traffic engineering, this chapter first estimates the possible traffic of the satellite network. We assume that the satellite network flow is a stationary random process, and predict the possible traffic of the satellite network of a certain time in the future using the stationary time series forecasting method according to the traffic of the terrestrial network, the traffic history of the satellite network, and the traffic global distribution of the satellite network. And then, we use the traffic engineering method with a punishment envelope to allocate the traffic for the satellite network and make a simulation.

4.2 Traffic Prediction of the Satellite Network

In order to make the satellite network system adapt to the traffic changes better, we predict the traffic in the satellite network. The traffic prediction needs the previous traffic matrix of the satellite network, which is not provided in the existing literatures. The satellite network will become the bypass network of the terrestrial network, and may take on the same or similar traffic to that of the terrestrial network. The traffic of the terrestrial network is an important and useful reference to the satellite network. In this part, we use a terrestrial traffic model to predict the traffic for the satellite network so as to provide a method for the future work of satellite network traffic prediction.

Although the traffic demand matrix of the satellite network is not provided in the previous literature studies, [17] provides the voice traffic that was expected to be taken on by the LEO satellite network system in 2005. As shown in Fig. 4.1, the world is divided into 12×24 cells; each cell occupies 15° longitude and 15° latitude. The traffic demand level of each cell corresponds to the communication time of this cell in one year. The communication time includes the communication time of the cell as a calling party and a called party.

This chapter uses hour as the basic unit of the prediction. The users have different traffic demands in different periods of time. [18] provides the basic laws of the traffic demands of satellite network users in 24 h as shown in Fig. 4.2. The time shown in Fig. 4.2 is the local time of users. We assume that this law is fit for all of the global users and the users' local time equals the solar time of their own region.

The proportions of satellite network traffic on all continents are provided by [19] as shown in Table 4.1. Table 4.1 shows the proportion of the network traffic among the continents in 2005, which is a statistical value of the average usage of all users on the continent.

Another important data source reference in this chapter is the United States education network (Abilene). The traffic data of Abilene was used in [19]. Although the detailed traffic data cannot be obtained, we can assume the traffic model according to the description in [19]. It is assumed that there is a traffic matrix including the traffic of some colleges and universities as well as some corporations of the United States from March 1, 2004 to September 10, 2004. The data sampling time is 5 min.

Intensity level and Corresponding Expected Traffic (2005)

Intensity level	1	2	3	4	5	6	7	8
Traffic (million minutes/year)	1.6	6.4	16	32	95	191	239	318

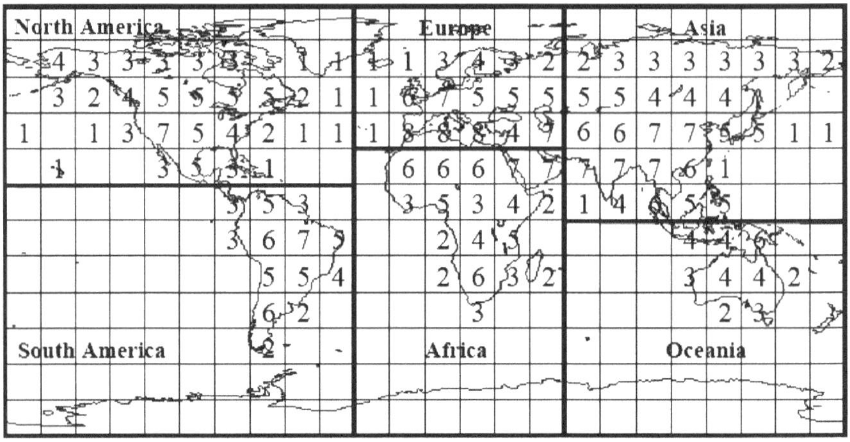

North America
Europe
Asia
South America
Africa
Oceania

Fig. 4.1 The voice traffic expected to be taken on by the LEO satellite network system in 2005

12
10
8
6
4
2
Percentage /%
0
5
10
15
20
25
Hour

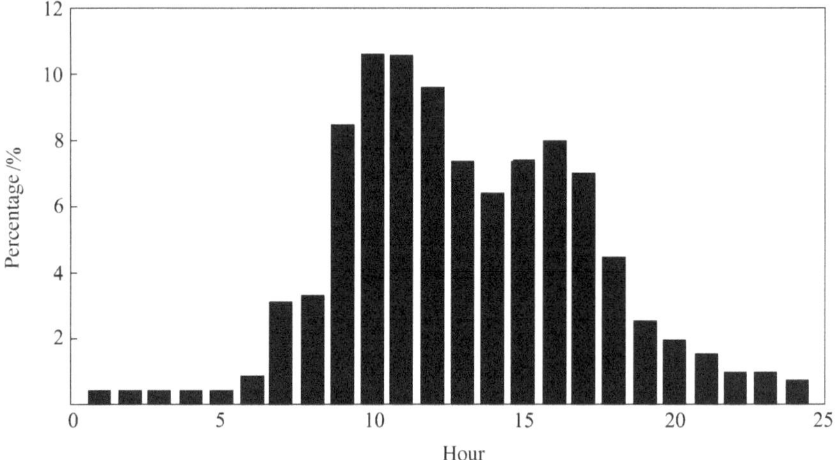

Fig. 4.2 The laws of the traffic demands of satellite network users in 24 h

The prediction of the satellite network traffic can be made through the above data. First, we will number each cell in Fig. 4.1. Notice that not all the cells need to be numbered. In order to reduce the computation overhead, we only need to number the cells that have traffic demand. The cells are numbered from left to right and from top to bottom. The cell ID is like 1,..., M, where $M = 120$.

Table 4.1 The proportion of the network traffic among the continent

Source	Destination					
	North America	Europe	Asia	South America	Africa	Oceania
North America	86.18	6.74	4.18	1.76	0.45	0.70
Europe	25.10	55.88	13.52	1.62	2.84	1.04
Asia	24.04	20.89	47.74	1.15	1.75	4.43
South America	52.39	13.02	5.96	25.12	1.85	1.66
Africa	25.63	43.34	17.33	3.53	7.95	2.22
Oceania	26.48	10.58	29.22	2.11	1.49	30.12

Since only the traffic duration is provided by [2, 17, 18], we use the traffic duration on behalf of the traffic demand. Figure 4.1 provides the annual traffic demand of the year 2005. Assuming that the traffic demand distribution is the same each and everyday of the year, the traffic of each cell for each day is hence equal to the annual traffic/365. The data provided by [17] only represents the global traffic distribution. We assume that the same daily traffic demand does not affect the following traffic estimation. Data in Fig. 4.2 provides the traffic distribution at different times of the day.

Assume that X_i is the daily traffic of the cell i, and $\kappa(t)$ is the function of time t and traffic percentage. For example, when $t = 7$, $\kappa(7) = 3.5\ \%$ according to Fig. 4.2. $\chi_i(t)$ is the traffic of cell i in time t. Obviously, $\chi_i(t) = X_i \times \kappa(t)$. $\varepsilon_{ij}(t)$ is the traffic from cell i to cell j during time t. Based on the above assumption, we have

$$\begin{cases} \sum_{j=1}^{M} \varepsilon_{1j}(t) + \sum_{i=1}^{M} \varepsilon_{i1}(t) = \chi_1(t) \\ \quad\cdot \qquad \cdot \qquad \cdot \\ \quad\cdot \qquad \cdot \qquad \cdot \\ \quad\cdot \qquad \cdot \qquad \cdot \\ \sum_{j=1}^{M} \varepsilon_{Mj}(t) + \sum_{i=1}^{M} \varepsilon_{iM}(t) = \chi_M(t) \end{cases} \tag{4.1}$$

The simultaneous equations (4.1) mean that the total traffic of a cell during time t is equal to its outgoing traffic and incoming traffic put together. There are M^2 number of variables in the simultaneous equations (4.1), and only M number of equations.

Considering the traffic proportion provided by Table 4.1, we assume that the total of the outgoing traffic of North America, Europe, Asia, South America, Africa, and Oceania are $v_1 \sim v_6$ separately. The number of cells in these continents are $n_1 \sim n_6$ separately. $N_1 \sim N_6$ are the set of cells belonging to these continents. φ_{xy} is the proportion of traffic from continent x to continent y. It is easy to know from Fig. 4.1 that $n_1 = 31$, $n_2 = 18$, $n_3 = 31$, $n_4 = 13$, $n_5 = 18$, $n_6 = 9$. If the traffic is distributed evenly, then $\forall \varepsilon_{ij}$, $i \in N_p, j \in N_q$, we have $\varepsilon_{ij} = v_p \cdot \varphi_{pq}/n_p \cdot n_q$. Therefore, variable ε_{ij} in the simultaneous equations (4.1) can be transformed into $v_p \times \varphi_{pq}$, where φ_{pq} is known. The equations (4.1) are hence transformed into

$$\begin{cases} \sum_{j=1}^{M} v_{I(1)} \cdot \varphi_{I(1)I(j)}/n_{I(1)} \cdot n_{I(j)} + \sum_{j=1}^{M} v_{I(j)} \cdot \varphi_{I(j)I(1)}/n_{I(j)} \cdot n_{I(1)} = \chi_1 \\ \cdot \quad \cdot \quad \cdot \\ \sum_{j=1}^{M} v_{I(i)} \cdot \varphi_{I(i)I(j)}/n_{I(i)} \cdot n_{I(j)} + \sum_{j=1}^{M} v_{I(j)} \cdot \varphi_{I(j)I(i)}/n_{I(j)} \cdot n_{I(i)} = \chi_i \\ \cdot \quad \cdot \quad \cdot \\ \sum_{j=1}^{M} v_{I(M)} \cdot \varphi_{I(M)I(j)}/n_{I(M)} \cdot n_{I(j)} + \sum_{j=1}^{M} v_{I(j)} \cdot \varphi_{I(j)I(M)}/n_{I(j)} \cdot n_{I(M)} = \chi_M \end{cases} \tag{4.2}$$

where $i \in N_{I(i)}$.

Thus, the M^2 number of variables in (4.2) are transformed into $v_1 \sim v_6$ six variables, and the number of equations is unchanged ($M = 120$). Split the equations in (4.2) into 20 equations with a full rank coefficient matrix, then we can get 20 groups of solutions of $v_1 \sim v_6$. Assume that $\overline{v_1} - \overline{v_6}$ are the arithmetic means of $v_1 \sim v_6$, the proportion of traffic demand, *td*, among these continents can be got, and the traffic ratio $DP_{x_1y_1-x_2y_2}$ can be represented as follows:

$$DP_{x_1y_1-x_2y_2} = \frac{td_{x_1y_1}}{td_{x_2y_2}} = \frac{v_{x_1}\varphi_{x_1y_1}}{v_{x_2}\varphi_{x_2y_2}} \tag{4.3}$$

Assume the traffic is evenly distributed when finding the solution of $v_1 \sim v_6$. The traffic demand of a certain pair of cells can be estimated by its traffic density level. The traffic density level of cell i is marked as $IL(i)$. The traffic demand td_{ij} from cell i to cell j is hence as follows:

$$td_{ij} = v_{I(i)} \cdot \varphi_{I(i)I(j)} \frac{IL(i) \cdot IL(j)}{\sum_{p \in N_{I(i)}, q \in N_{I(j)}} IL(p) \cdot IL(q)} \qquad i \in N_{I(i)}, j \in N_{I(j)} \tag{4.4}$$

Then, let's consider the traffic of Abilene. First, we process the basic assumption of Abilene traffic. The 5 min sampling time is combined as 1 hour sampling time which is marked as $\Delta_{ij}(t)$, where i and j are the cell ID located in the United States, t is the sampling time. Take the communication between central and eastern United States (the two zones in the bold wireframe in Fig. 4.3) as an example. The central and eastern United States is further divided into 4 sub-regions according to the longitude and latitude since the traffic between central and eastern U.S. is still too large. As shown in the bold wire frame in Fig. 4.3, the sub-regions IDs 1–4 are located in central US and 5–8 are located in eastern U.S. Take the traffic from sub-region 1 to sub-region 6 as an example. The traffic between these two sub-regions from March 1, 2004 to April 1, 2004 is a time series S_t, which can represent the traffic between central and eastern U.S. Since the growth of the traffic demand of the Abilene network is slow [19], it can be deemed as no systematically changes or no trend of the mathematical expectation of traffic demand in one year. The variance of the traffic demand can also be regarded as no systematically changes. The traffic demand data in one year can reflect the changing cycle of the traffic in the long term. Therefore, S_t is a stationary time series according to the definition in [20].

Assume the traffic demand from sub-region 1 to sub-region 6 is $td_{ij}(t)$, $\{td_{ij}(1), \ldots, td_{ij}(n)\}$ is used to predict the stationary series S_t:$\{td_{ij}(t), t \geq n+1\}$. The Durbin–Levinson algorithm is used in this chapter to do the step forecast of traffic demand, the result is marked as $\overline{\overline{td_{ij}}}(n+1)$.

First, a zero-mean time series $td'_{ij}(t)$, $td'_{ij}(t) = td_{ij}(t) - E(td_{ij}(t))$ is constructed. Then, $\overline{\overline{td'_{ij}}}(n+1)$ is presented as $\overline{\overline{td'_{ij}}}(n+1) = \phi_{n1} td'_{ij}(n) + \ldots + \phi_{nn} td'_{ij}(1)$, $n \geq 1$; and the mean square error ψ_n is presented as $\psi_n = E(td'_{ij}(n+1) - \overline{\overline{td'_{ij}}}(n+1))^2$, $n \geq 1$. Assume that the auto covariance function of $td'_{ij}(t)$ is $\gamma(\cdot)$, the ϕ_{nj}, and ψ_n can hence be calculated recursively as follows:

(1) $\phi_{11} = \frac{\gamma(1)}{\gamma(0)}, \quad \psi_0 = \gamma(0),$

(2) $\phi_{nn} = [\gamma(n) - \sum_{j=1}^{n-1} \phi_{n-1,j}\gamma(n-j)]\psi_{n-1}^{-1}$

(3) $\begin{bmatrix} \phi_{n1} \\ \cdot \\ \cdot \\ \cdot \\ \phi_{n,n-1} \end{bmatrix} = \begin{bmatrix} \phi_{n-1,1} \\ \cdot \\ \cdot \\ \cdot \\ \phi_{n,n-1} \end{bmatrix} - \phi_{nn} \begin{bmatrix} \phi_{n-1,n-1} \\ \cdot \\ \cdot \\ \cdot \\ \phi_{n,n-1} \end{bmatrix}$

(4) $\psi_n = \psi_{n-1}[1 - \psi_{nn}^2]$

The proof can be found in [21]. After $\overline{\overline{td'_{ij}}}(n+1)$ is got, $\overline{\overline{td_{ij}}}(n+1) = \overline{\overline{td'_{ij}}}(n+1) + E(td_{ij}(t))$ can be known. The above method gives a forecast of the traffic demand between central and Eastern America. When $h > 1$, the forecast can be done similarly. The traffic of a cell in anyplace of the world can be predicted through the prediction of traffic between central and Eastern America. Assume that the ID of Central America is i, and Eastern America is j, the traffic demand in time $n+1$ between them is predicted as $\overline{\overline{td_{ij}}}(n+1)$, then the traffic between any cells p and q can be calculated as:

$$\overline{\overline{td_{pq}}}(n+1) = \frac{td_{pq}}{td_{ij}} \overline{\overline{td_{ij}}}(n+1) \tag{4.5}$$

Fig. 4.3 Traffic between central and eastern U.S.

4.3 Traffic Engineering of the Satellite Network

Although the satellite network traffic prediction is given in Sect. 4.2, it is only limited to the normal traffic demand. It is said in [19] that there exist some unpredicted traffic spikes as shown in Fig. 4.4. When these traffic spikes appear, the network performance using the traditional traffic engineering methods, such as oblivious routing or prediction-based traffic engineering, will be affected greatly as some link utilizations will increase sharply, resulting in link stoppage.

So far, these traffic spikes have not been recorded in any documents. However, the satellite network is an important bypass network of the terrestrial network; traffic engineering of the satellite network is needed to prevent the impact of traffic spikes. Since the onboard resources and backup routers of the satellite network are limited, it is more sensitive to the impact of traffic spikes; robust traffic engineering with less overhead is more necessary for the satellite network.

Prediction-based traffic engineering needs to collect the previous traffic demands and uses them as the basis for traffic engineering. When the network traffic deviates a lot from the collected traffic demand, the performance of prediction-based traffic engineering will deteriorate sharply. Hence it is not robust traffic engineering. The essence of oblivious routing is to consider all the possible traffic demand which leads to great computational overhead, and is not conducive to the optimization for the normal traffic.

This chapter draws on the advantages of prediction-based traffic engineering and oblivious routing, and uses a novel traffic engineering method, i.e., prediction-based traffic engineering with a cost envelope. The method is first proposed in [19] for the terrestrial network. Its basic idea is to optimize for the Predicted Traffic demand within a Cost Envelope (PTCE).

4.3.1 Concepts and Definitions

Traffic engineering of the satellite network is based on the NSGRP in Chap. 3. After the snapshot is divided in NSGRP, traffic engineering is implemented in each snapshot. The main objective of traffic engineering is to balance the traffic of the network to prevent congestion. The precursor of the network congestion is that some link utility is greater than a certain threshold. Therefore, the optimization objective of traffic engineering from the network's aspect is the Maximum Link

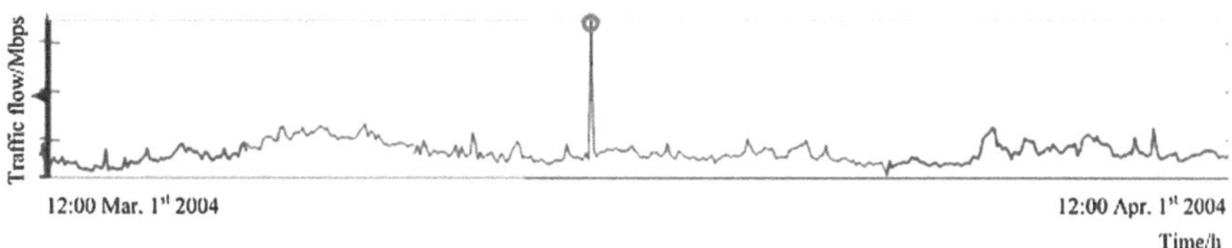

Fig. 4.4 The traffic spike in March 2004

Utility (MLU) of the network. Of course, there are also some literatures such as [7, 8] that use network overhead as the optimization objective. In this chapter, MLU is used as the optimization objective.

We model the topology of each snapshot divided by NSGRP as a directed graph $G(V, E)$, where V is the set of onboard routers and E is the set of IOLs and ISLs. Notice that since the propagation delay of IOLs between the MEO-GEO layer and the LEO-GEO layer is much larger than the ISLs in the LEO and MEO layers and the IOLs between the LEO and MEO layers, they are used as backup links and are not involved in traffic engineering. The basic concepts used in the PTCE will be defined as follows.

Definition 4.1 (*Traffic demand matrix, TD*) The traffic demand matrix TD is a $m \times m$ matrix. Each element td_{ab} in matrix $TD_{m \times m}$ is the traffic demand from the LEO satellite a to the LEO satellite b with the unit Mbps. m is the number of LEO satellites.

The traffic demand matrix only considers the traffic demand between LEO and LEO satellites. This is because the LEO satellites are used as the entrance and exit satellites in the Tr constellation, which are equivalent to the border gateways of the terrestrial AS. The traffic demand between each pair of terrestrial OD gateways can be expressed as the traffic demand between the pair of LEO satellites bound with the OD gateways. Therefore, the matrix $TD_{m \times m}$ is also the traffic demand of the whole satellite network, and is the input of PTCE.

Definition 4.2 (*Traffic engineering scheme P*) The traffic engineering scheme P is a three-dimensional matrix $Pm \times m \times n$. Each element $p_{ab}(i, j)$ of matrix $Pm \times m \times n$ is the proportion of the traffic demand td_{ab} allocated on link (i, j), which is called a flow allocation value of the traffic engineering scheme P. n is the total number of ISLs and IOLs in the LEO/MEO layers of the Tr constellation.

The method used in this chapter is traffic engineering based on links. It can be transformed to traffic engineering based on paths according to [22] in practice, and also can be transformed to an equal weighted division traffic engineering of OSPF according to [11]. $p_{ab}(i, j) = 0$ indicates that the traffic demand td_{ab} from a to b will not pass through link (i, j), while $p_{ab}(i, j) = 1$ indicates that the traffic demand td_{ab} will pass through link (i, j) totally.

Definition 4.3 (*Maximum link utility, MLU*) The MLU of P under the traffic demand matrix TD is the maximum link utility of all the links in the LEO/MEO layers of the Tr constellation under the traffic matrix TD when using the traffic engineering scheme P, which can be presented as

$$MLU(P, TD) = \max \sum_{(i,j) \in E \cap a,b \in V} td_{ab} \cdot p_{ab}(i,j)/c(i,j) \tag{4.6}$$

where $c(i, j)$ is the bandwidth of the link (i, j).

In Eq. (4.6), $td_{ab} \cdot p_{ab}(i, j)$ indicates the real traffic that passes through the link (i, j) of the traffic demand td_{ab}.

Definition 4.4 (*Optimum traffic engineering value, OP*) An optimum traffic engineering value of a given traffic demand matrix *TD* is the MLU of the network made by the best traffic engineering scheme P_{optimal} under the traffic matrix *TD*, which can be presented as

$$OP(TD) = \min_{P \in \Lambda} MLU(P, TD) \tag{4.7}$$

where Λ is the set of all possible traffic engineering schemes.

In Eq. (4.7), the optimum traffic engineering scheme may be more than one. In this case, we can choose the solution that can make the total path delay between all the OD satellites or the variance of the link utilities of the network minimum. Moreover, we can give weights to these two values according to the application scenario, and choose the optimal solution that has the minimum weighted average.

Definition 4.5 (*Optimal performance ratio, PR*) The optimal performance ratio of a given traffic engineering scheme *P* under a traffic demand matrix *TD* is defined as the MLU made by *P* under the traffic matrix *TD* divided by the optimum traffic engineering value under *TD*, which can be presented as

$$PR(P, TD) = MLU(P, TD)/OP(TD) \tag{4.8}$$

where $PR(P, TD) \geq 1$.

In Eq. (4.8), $PR(P, TD)$ indicates the degree of optimization of the traffic engineering scheme *P*. $PR(P, TD) = 1$ means that *P* is optimum. A large value of *PR* means a low degree of optimization.

Since the traffic of the satellite network changes over time, a single traffic matrix cannot reflect the real traffic of the network. So the inputs of the prediction-based traffic engineering scheme and oblivious routing are sets of traffic demand matrices. Therefore, the definitions of MLU and *PR* need to be extended in order to allocate the traffic in practice. Assume that ***TD*** is the set of traffic demand matrix *TD*. The definition of MLU can be extended as follows.

Definition 4.6 (*Maximum link utility,* **MLU**) The maximum link utility **MLU** of *P* under the set of traffic demand matrix ***TD*** is the maximum value of MLU of *P* under each $TD \in \boldsymbol{TD}$, which can be presented as

$$\mathbf{MLU}(P, \boldsymbol{TD}) = \max_{TD \in \boldsymbol{TD}} \text{MLU}(P, TD) \tag{4.9}$$

The traffic engineering scheme *P* that makes the **MLU** under ***TD*** minimum is called the optimum traffic engineering scheme of ***TD***. The **MLU** is called the optimum MLU of the set ***TD***. Similarly, we extend the definition of *PR* as follow.

Definition 4.7 (*Optimal performance ratio,* **PR**) The optimal performance ratio ***PR*** of *P* under the set of traffic demand matrix ***TD*** is the maximum *PR* of *P* under each *TD*∈ ***TD***, which can be presented as

$$PR(P, \mathbf{TD}) = \max_{TD \in \mathbf{TD}} PR(P, TD) \quad (4.10)$$

Definition 4.8 (*Cost envelope,* $\Re$) If the MLU or *OP* of *P* under all the possible *TD*s is no more than $\Re$, it is called that *P* has the cost envelope.

Before introducing the implementation of the satellite network traffic engineering, we need to make some constraints of the traffic engineering scheme in order to meet the actual situation. In the actual satellite network, the traffic engineering scheme *P* needs to meet the following constraints.

$$\begin{cases} 0 \le p_{ab}(i,j) \le 1 \quad \forall (i,j) \in E \\ \sum_{(i,j)\in E} p_{ab}(i,j) - \sum_{(j,i)\in E} p_{ab}(j,i) = 0 \quad \forall a \ne b, \forall i \ne a, b \\ \sum_{(a,j)\in E} p_{ab}(a,j) - \sum_{(j,a)\in E} p_{ab}(j,a) = 1 \quad \forall a \ne b \\ \sum_{a,b\in V} td_{ab} \cdot p_{ab}(i,j) \le c(i,j) \quad \forall (i,j) \in E \end{cases} \quad (4.11)$$

The constrains in Eq. (4.11) represent the non-negative value of flow distribution; the input traffic of an OD satellites in a satellite node is equal to the output traffic of this OD satellites in this node; the flow distribution ratio on the output links of the source node is 1; and the allocation traffic on link (*i*, *j*) of all the OD satellites cannot be greater than its link capacity.

4.3.2 The Implementation of PTCE

The prediction-based traffic engineering scheme uses a set of traffic demands to predict and optimize for the future traffic of the network. Given a set of traffic demands $TD = \{d_{ij}(1), \ldots, d_{ij}(n)\}$ as mentioned in Sect. 4.2, the convex combination of the elements in *TD* can be used to plan for the future traffic, a typical application is the convex hull of *TD*. Assume that ***TD*** is the convex hull of *TD*, the elements of ***TD*** $\mathbf{td}_{ij} = \sum a_k d_{ij}(k)$, where $0 \le a_k \le 1$ and $\sum a_k = 1$. In this way, the problem of traffic engineering for a certain time in the future is converted into the problem of traffic engineering for the set of ***TD***.

If the future traffic falls into the set ***TD***, the prediction-based traffic engineering scheme will be quite efficient. However, if the future traffic falls outside the set of ***TD***, the performance may deteriorate sharply. Notice that for the traffic outside the convex hull of the set of ***TD***, the traffic prediction method mentioned in Sect. 4.2 does not work. Of course, we can extend the convex hull of *TD* to make it contain more traffic demand matrices, such as let $a_k \le 0$ or $a_k \ge 1$. Although this can alleviate the influence of the traffic change on the performance of the algorithm to some degree, the algorithm overhead increases dramatically with the expansion of the convex hull. A solution of this problem is to separate the optimization with the expansion of the convex hull.

The method of PTCE is to use the cost envelope $\Re$ as the performance limitations of all the traffic engineering schemes. The ***PR*** and **MLU** of all the traffic engineering schemes under all the possible traffic matrices must be within this cost envelope, and the optimization is done for the set of normal traffic demand matrices. In this way, the network performance under the traffic spike will not exceed the cost envelope, and it will be close to the optimum when the traffic is normal. Assume that ***TD*** is the set of normal traffic, and $\aleph$ is the set of all possible traffic demands. The model of PTCE can be abstracted as

$$\min \mathbf{MLU}(P, TD)$$

s. t. P is a traffic engineering scheme that satisfies the constraints in (4.11);

$$\mathrm{MLU}(P, TD) \leq \Re \, \forall \, TD \in \aleph.$$

The basic framework of PTCE is shown in Fig. 4.5.

For the cost envelope, we can choose the minimum **MLU** of the traffic engineering scheme P under all the possible traffic matrices multiplied by a coefficient. That is

$$\Re = \lambda \min \mathbf{MLU}(P, \aleph) \tag{4.12}$$

where λ is the scaling factor. If λ is large, the algorithm can handle the unpredictable traffic spike well, but performs poorly when the traffic is normal. On the contrary, if λ is small, the algorithm can handle the normal traffic well, but performs poorly when the traffic spike is coming, and is lack of robustness.

According to the basic idea of PTCE, it can be modeled as follows.

$$\min \mathbf{MLU}(P, \mathbf{TD}) \tag{4.13}$$

s.t. P satisfies the constraints of (4.11);

$$\begin{gathered} \forall \operatorname{link}(i,j), \;\; \forall td_{ab} \geq 0: \\ \sum\nolimits_{b \in V} td_{ab} \leq \sum\nolimits_{j \in V,(a,j) \in E} c(a,j); \\ \sum\nolimits_{b \in V} td_{ba} \leq \sum\nolimits_{i \in V,(i,a) \in E} c(i,a); \\ \sum_{a,b \in V} td_{ab} \cdot p_{ab}(i,j)/c(i,j) \leq \Re \end{gathered} \tag{4.14}$$

Obviously, constraints (4.14) are not a standard form of linear programming. However, it can be solved by finding the solution of the following linear programming and testing if the solution is less than or equal to $\Re$.

$$\max \sum_{a,b \in V} td_{ab} \cdot p_{ab}(i,j)/c(i,j) \tag{4.15}$$

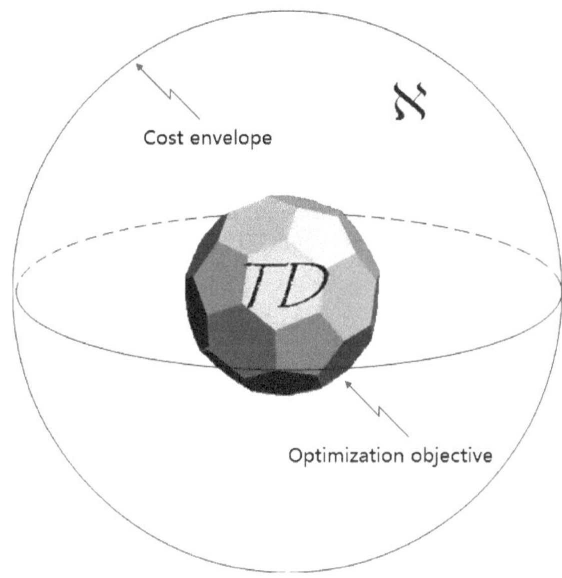

Fig. 4.5 The framework of PTCE

$$\text{s.t.}\quad td_{ab} \geq 0 \quad \forall a, b \in V$$

$$\sum\nolimits_{b \in V} td_{ab} \leq \sum\nolimits_{j \in V,(a,j) \in E} c(a,j)$$

$$\sum_{b \in V} td_{ba} \leq \sum_{i \in V,(i,a) \in E} c(i,a)$$

Using the linear programming duality theorem, we know that if the objective function in (4.15) is less than or equal to $\Re$, if and only if the following constraints are satisfied.

$$\begin{aligned}
&\alpha_{(i,j)}(a) \geq 0, \quad \beta_{(i,j)}(a) \geq 0 \quad \forall a \in V; \\
&p_{ab}(i,j)/c(i,j) \leq \alpha_{(i,j)}(a) + \beta_{(i,j)}(b) \quad \forall a, b \in V; \\
&\sum\nolimits_{a \in V} \left(\alpha_{(i,j)}(a) \cdot \sum\nolimits_{j \in V,(a,j) \in E} c(a,j) + \beta_{(i,j)}(a) \cdot \sum\nolimits_{i \in V,(i,a) \in E} c(i,a)\right) \leq \Re
\end{aligned} \tag{4.16}$$

In which, $\alpha_{(i,j)}(a)$ and $\beta_{(i,j)}(a)$ are the dual multipliers of the node capacity constraints $\sum_{j \in V(a,j) \in E} c(a, j)$ and $\sum_{i \in V(i,a) \in E} c(i, a)$, respectively. So, we can use (4.16) as the constraints instead of (4.14).

In this way, we can get a traffic engineering method with less overhead, whose objective function is **MLU**, and the constraint is $\Re$. It can optimize for the normal traffic set ***TD***. ***PR*** can also be used as the objective function, and the implementation is similar to the method using **MLU** as the objective function.

For a certain period of time *n*, the normal traffic matrix set ***TD*** can be got by the following steps.

(1) Choose the traffic demands of sub-regions 1–4 and 5–8 separately at $t = 1, \ldots, n-1$, and use the Durbin–Levinson algorithm described in (4.2) to predict the traffic for time $t = n$. Then we can get the traffic demand between 32 pairs of sub-regions marked as $d_{xy}^{(k)}(n)$, where x and y is the number of sub-regions, and $k = 1, \ldots, 32$. Substitute the cell number of x, y for x, y, we have $d_{ij}^{(k)}(n)$, $k = 1, \ldots, 32$.

(2) Use the 32 traffic demands as the source data, and calculate the traffic demand matrices of the $M \times M$ OD cells all over the world using Eq. (4.5). We can get $\boldsymbol{TD} = \{TD_1, \ldots, TD_{32}\}$

4.4 Simulations and Results

We choose the *Tr* constellation as the experimental subject, and test the network performance of the *Tr* system in one day. For ease of computing, we use the U.S. Eastern time as the system time, the times mentioned below all refer to the U.S. Eastern time. The traffic model mentioned in Sect. 4.2 in March 16, 2004 is used as the data source. The source data is mapped to each pair of cells all over the world through Eq. (4.5). The results are used as the actual traffic of the satellite network in March 16, 2004. Assume that the traffic spike shown in Fig. 4.4 just happened in between the east and west of the United States between 21:00 March 16 to 0:00 March 17, 2004, and the traffic remained otherwise normal all the time.

The traffic data of the model between March 1, 2004 and March 15, 2004 is used as the history data of traffic demands, as shown in Fig. 4.6.

In Fig. 4.6, $TD_x^{(y)}(z)$ represents the network traffic matrix of the xth group of data in time z of the yth day. Each group of traffic data is formed by each pair of sub-regions in 15 days. The $\boldsymbol{TD}$ of the satellite network on March 16 can be got using the following method.

The $\boldsymbol{TD}$ of the whole network in time k of March 16

$$
\begin{array}{l}
TD_1^{(1)}(k),\ TD_1^{(2)}(k),\ \ldots, TD_1^{(15)}(k) \overset{Durbin-Levinson}{\longrightarrow} \overline{\overline{\mathrm{TD}}}_1^{(16)}(k) \\
TD_2^{(1)}(k),\ TD_2^{(2)}(k),\ \ldots, TD_2^{(15)}(k) \overset{Durbin-Levinson}{\longrightarrow} \overline{\overline{\mathrm{TD}}}_2^{(16)}(k) \\
\quad\vdots \qquad\qquad \vdots \qquad\qquad\qquad \vdots \qquad\qquad\qquad\qquad\qquad \vdots \\
TD_{32}^{(1)}(k),\ TD_{32}^{(2)}(k),\ \ldots, TD_{32}^{(15)}(k) \overset{Durbin-Levinson}{\longrightarrow} \overline{\overline{\mathrm{TD}}}_{32}^{(16)}(k)
\end{array}
$$

$$
\boldsymbol{TD} = \{\overline{\overline{TD}}_1^{(16)}(k), \overline{\overline{TD}}_2^{(16)}(k), \ldots, \overline{\overline{TD}}_{32}^{(16)}(k)\}
$$

In order to verify the performance of PTCE in the *Tr* satellite constellation, we compare it with traditional oblivious routing and the prediction-based traffic engineering scheme. Oblivious routing uses the minimum convex set that contains

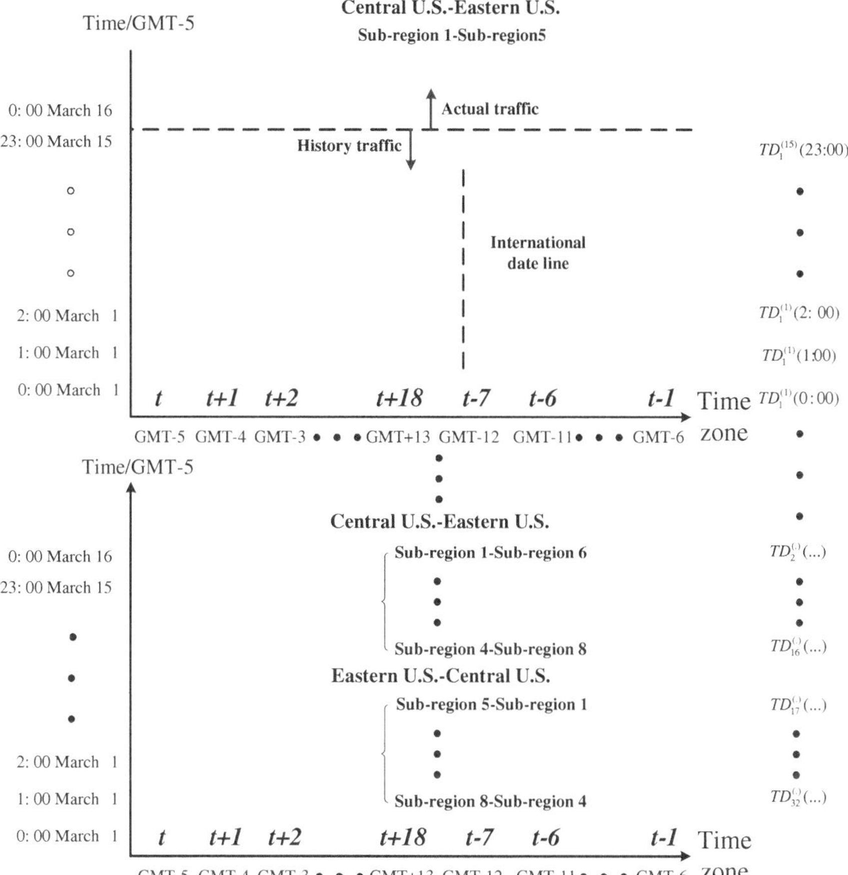

Fig. 4.6 The acquirement of the satellite network traffic data

the traffic spike as the optimization object, and the prediction-based traffic engineering scheme optimizes for the ***TD*** traffic demands. The whole satellite network uses these three traffic engineering methods on March 16, and the ***PR***, **MLU** are shown in Figs. 4.7 and 4.8, respectively.

The ***PR***, **MLU** and packet loss rate of the PTCE, prediction-based traffic engineering and oblivious routing under the unpredicted traffic spike are shown in Figs. 4.7, 4.8, and 4.9. As shown in Fig. 4.7, the performance of the prediction-based traffic engineering method deteriorates sharply in ***PR*** and deviates a lot from the optimum solution when the traffic spike happens at 21:00, March 16. Although oblivious routing performs well in ***PR*** when the traffic spike comes, it deviates a lot from the optimum solution when the traffic is normal for its optimization set is

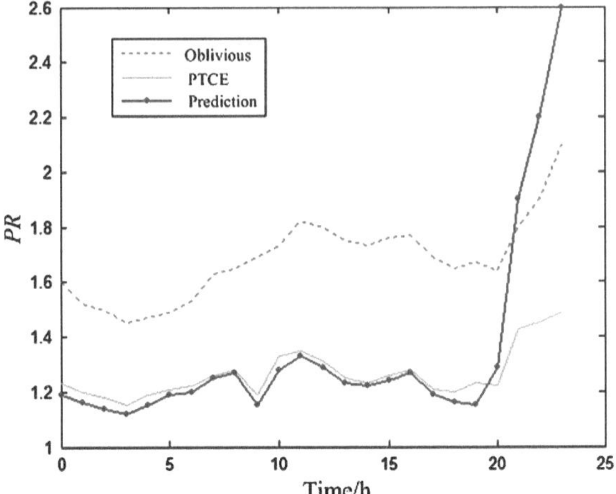

Fig. 4.7 The *PR* comparison of three traffic engineering methods

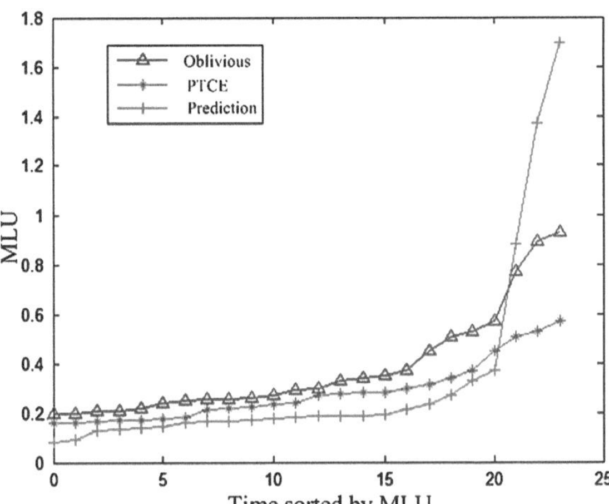

Fig. 4.8 The MLU comparisons of three traffic engineering methods

too large. The PTCE method considers both the normal traffic and traffic spike, which makes its ***PR*** close to the optimum solution when the traffic is normal. It does not deviate too much from the optimum solution when the traffic spike comes. The situation shown in Figs. 4.8 and 4.9 is similar to the situation in Fig. 4.7. In order to show the trends of performance, the time is sorted by the MLU and packet loss rate.

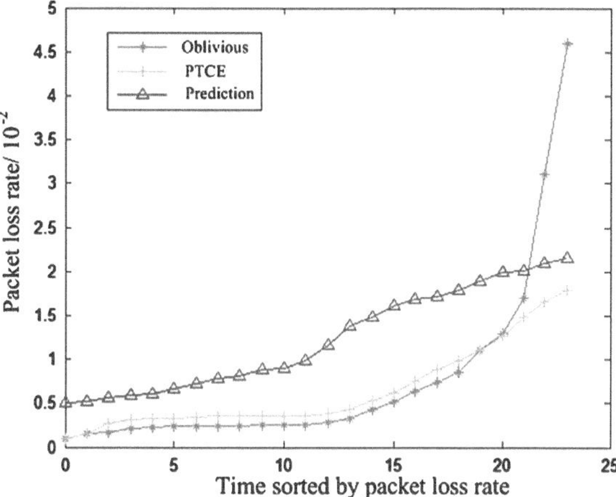

Fig. 4.9 The packet loss rate comparison of three traffic engineering methods

4.5 Summary

This chapter first predicts the traffic for the satellite network using the Durbin–Levinson algorithm, a time series analysis method, according to the traffic density distribution of the satellite network, the traffic-time distribution of the satellite network, and a terrestrial traffic model. And then, we use a new traffic engineering method called PTCE proposed first in [13] to optimize the traffic for the satellite network. PTCE uses the normal traffic of the satellite network as the optimization object while setting a cost envelope for the unpredicted traffic spike. In this way, the traffic engineering scheme can be close to optimum solution when the traffic is normal while not deteriorating a lot when a traffic spike comes. The simulation results show that the performance of PTCE is close to prediction-based traffic engineering when the traffic is normal, and it is better than oblivious routing when the traffic spike comes. Therefore, it is a traffic engineering method fit for satellite networks.

References

1. Agarwal S, Nucci A, Bhattacharyya S (2005) Measuring the shared fate of IGP engineering and inter-domain traffic. In: Proceedings of the 13th international conference on network protocols (ICNP'05), vol 11. Boston, pp 30–39
2. Wang H (2008) Efficient and robust traffic engineering in a dynamic environment. Ph. D. thesis, Yale University, New Haven
3. Applegate D, Breslau L, Cohen E (2004) Coping with network failures: Routing strategies for optimal demand oblivious restoration. In: Proceedings of joint international conference on measurement and modeling of computer systems (SIGMETRICS), vol 6. New York, pp 243–249

4. Applegate D, Cohen E (2003) Making intra-domain routing robust to changing and uncertain traffic demands: understanding fundamental tradeoffs. In: Proceedings of ACM SIGCOMM'03, vol 8. Karlsruhe, pp 313–324
5. Elwalid A, Jin C, Low S et al (2001) MATE: MPLS adaptive traffic engineering. In: Proceedings of IEEE INFOCOM'01, vol 4. Anchorage, pp 453–464
6. Feamster N, Winick J, Rexford J (2004) A model of BGP routing for network engineering. In: Proceedings of joint international conference on measurement and modeling of computer systems (SIGMETRICS), vol 6. New York, pp 331–342
7. Fortz B, Rexford J, Thorup M (2002) Traffic engineering with traditional IP routing protocols. IEEE Commun Mag 10:118–124
8. Fortz B, Thorup M (2000) Internet traffic engineering by optimizing OSPF weights. In: Proceedings of IEEE INFOCOM'00, vol 3. Tel Aviv, pp 519–528
9. Kandula S, Katabi D, Davie B et al (2005) Walking the tightrope: responsive yet stable traffic engineering. In: Proceeding of ACM SIGCOMM'05, vol 8. Philadelphia, pp 345–357
10. Kodialam M, Lakshman TV, Sengupta S (2004) Efficient and robust routing of highly variable traffic. In: Proceedings of third workshop on hot topics in networks (HotNets-III), vol 10. San Diego, pp 79–86
11. Sridharan A, Guerin R, Diot C (2003) Achieving near optimal traffic engineering solutions in current OSPF/ISIS networks. In: Proceedings of IEEE INFOCOM'03, vol 4. San Francisco, pp 234–247
12. Zhang SR, McKeown N (2004) Designing a predictable internet backbone network. In: Proceedings of third workshop on hot topics in networks (HotNets-III), vol 10. San Diego, pp 436–442
13. Wang H, Xie HY, Qiu LL et al (2006) COPE: traffic engineering in dynamic networks. In: Proceedings of ACM SIGCOMM'06, vol 10. Pisa Itlay, pp 99–110
14. Grover W, Tipper D (2005) Design and operation of survivable networks. J Netw Syst Manage 13(1):7–11
15. Markopoulou A, Iannaccone G, Bhattacharyya S et al (2004) Characterization of failures in an IP backbone network. In: Proceedings of IEEE INFOCOM'04, vol 4, no 4. Hong Kong, pp 2307–2317
16. Valiant LG (1982) A scheme for fast parallel communication. SIAM J Comput 11(7):350–361
17. Voilet MD (1995) The development and application of a cost per minute metric of the evaluation of mobile satellite systems in a limited-growth voice communications market. Master thesis, Massachusetts Institute of Technology, Cambridge
18. Perdigues J, Werner M, Karafolas N (2001) Methodology for traffic analysis and ISL capacity dimensioning in broadband satellite constellations using optical WDM networking. In: Proceedings of 19th AIAA international communication satellite systems conference (ICSSC'01), vol 4. Toulouse, pp 131–139
19. Chen C (2005) Advanced routing protocol for satellite and space networks. Ph. D. thesis, Georgia Institute of Technology, Atlanta
20. Fang ZG (1986) Introduction of time series analysis. Aerospace Press, Beijing (in Chinese)
21. Tian Z (2001) The theories and methods of time series. High Education Press, Beijing (In Chinese)
22. Ahuja R, Magnanti T, Orlin J (1993) Network Flows. Prentice Hall, New Jersey

Chapter 5
Satellite Network Multi QoS Objective Routing Algorithm

5.1 Introduction

In Chap. 4, we described satellite network traffic engineering from the aspect of the entire network to prevent network congestion. Satellite network traffic engineering can balance the load of the whole network and improve the operational efficiency of the satellite network system. However, it is not enough for providing the QoS services for the users. In recent years, the types and volume of business in the Internet have increased dramatically. Satellite networks, as an indispensible part of the global communication system, face the same problems as the Internet. The rapid increase of business types and volume requires satellite networks to be more reliable and provide more efficient service/ Moreover, a lot of new emerging business has QoS requirements. These QoS requirements come from the users' business needs. For example, some real-time business has a high requirement on the end-to-end delay. Some business, such as voice call, is sensitive to the delay jitters. And there are also some types of business that have high requirements on both delay and jitter. Therefore, how to meet the different QoS requirements of different users is a challenging problem.

Previous researches always focus on one single QoS requirement. For example, the delay jitters of the satellite network is improved in [1] from the aspect of flow switching. The switching of the satellite network is divided into two categories by Akyildiz et al.: Inter-satellite switching and ISL switching. The inter-satellite switching refers to the UDL link reconnection when a terrestrial gateway leaves its current entrance or exit satellite. The ISL switching refers to the path recalculation when some ISLs are no longer available with the movement of satellites. The delay jitter caused by flow switching is mitigated through location management and handover management in [1]. Location management is in charge of tracking and positioning of the user terminal for the access call, while handover management ensures the ongoing call is not affected by the changes of the service unit. The location management protocol processes the queuing and storage information in the local database while sending paging signals to locate the network users. The handover management analyzes the differences of user switching between a

F. Long, *Satellite Network Robust QoS-aware Routing*,
DOI: 10.1007/978-3-642-54353-1_5,

terrestrial network and a satellite network. It adopts a new switching management algorithm, hence improves the QoS of the network.

Similar to [1], the delay jitters of the satellite network were also improved from the aspect of link switching in [2]. The ATM-based satellite network was used as the research object in [2], and dynamic satellite network routing was also introduced in this chapter. The dynamic network topology is deemed as k periodically repeating snapshot series. Through the sliding window strategy, the source and destination nodes select a set that contains k optimal paths to minimize the link switching probability, thus to reduce the delay jitters. The disadvantage of this method is that it does not consider inter-satellite switching.

A prediction-based routing protocol to provide QoS service for the users of the satellite network was proposed in [3]. The protocol utilizes the predictability of the topology of the LEO-layered satellite to predict the traffic of an ISL in a short period in the future. At the same time, it calculates k optimal paths for each connection to maximize the minimum residual bandwidth. The actual routing paths are selected from the set of optimal paths to minimize link switching and balance the users' traffic. This protocol does not consider inter-satellite switching either. Besides, the dramatic computational overhead increase that comes along with the increase of k is another major defect.

The Probabilistic Routing Protocol, PRP utilizes the predictability of the movement of the LEO satellite network and the calling probability, and tries to reduce rerouting caused by link switching. If the switching probability of an ISL in the new call establishing phase is greater than a certain threshold value p, then this ISL will not be considered in the routing computation. The switching probability of an ISL is deduced by the regular hexagon footprint of the LEO satellite. This algorithm only considers the switching of the UDLs, hence has limited improvement on the delay jitters.

A QoS-based Routing Algorithm (QRA), for the satellite network was proposed by Chen C on the basis of the PRP [4]. QRA is committed to improve the shortcomings of the PRP. It reduces the delay jitters from the aspects of reducing the probabilities of inter-satellite switching and ISL switching. QRA is based on a general constellation model whose satellite footprints may be overlapped. Before routing computation, QRA predicts the possible switching of each link. Different from PRP, QRA does not delete the links that have a switching probability larger than p in the routing computation to prevent the increase of the new call blocking probability. In the rerouting part, QRA uses an improved rerouting protocol called Footprint Handover Rerouting Protocol, FHRP, to do fast rerouting for the users during link switches. Although QRA performs better than PRP in reducing the delay jitters, the increased algorithm complexity of QRA makes a larger protocol overhead. Besides, QRA is more likely to cause congestion. Since QRA is based on the original virtual topology grouping strategy which is very sensitive to protocol overhead, the robustness of the whole network routing system is not quite satisfactory.

Besides the delay jitter optimization satellite network routing algorithm, there are also some routing algorithms that optimize the ISL utilization. The FSA algorithm [5, 6] proposed by Chang et al. is representative of them all. The FSA routing algorithm solves the routing problem of the LEO satellite network according to the topology and the characteristic of satellite network traffic from the aspect of link assignment. The FSA algorithm divides the constellation cycle into a number of time slots. The network topology corresponding to each time slot is deemed as a state. This way, the dynamic topology of the satellite network is modeled as a finite state automaton. The routing problem of the LEO satellite network is thus transformed into the optimization problem of the link assignment for the network topology in the finite state automaton. The FSA algorithm uses the ISL utility as its only optimization target, which cannot guarantee the end-to-end delay for the users. In addition, the FSA algorithm lacks the measures to deal with the rerouting problem due to link switching. The link assignment is only for a certain traffic model which also makes the algorithm lack of scalability.

In order to reduce the communication overhead caused by the frequently updated topology, Tsai et al. proposed the Darting routing algorithm [7]. The algorithm updates routing only when it is necessary to reduce the communication overhead. The Darting algorithm utilizes a triggered routing update mechanism, and can be divided into forward updating and subsequent updating. Forward updating is responsible for updating the topology view held by the upstream node of current node in the routing path. When a node finds that the topology view held by the upstream node is different from its own, it will trigger the forward updating process. Subsequent updating is in charge of updating the topology view held by the downstream node of current node in the routing path. When the topology is changed, the satellite node will encapsulate the topology change information in the head of the data packets to be sent to inform its successor node about the topology changes. Both forward updating and subsequent updating are data-driven, i.e., when no data is sent, the topology information of each node remains unchanged. It is proved by the experiments that the Darting algorithm can reduce the communication overhead only when the network is with low load, but takes 72 % additional communication overhead [8] compared with the Bellman-Ford algorithm in [9] when the traffic load is high. The reason is that the routing updates frequently under high traffic load, which takes a lot of bandwidth to send the notice packet.

The end-to-end delay is another important QoS parameter besides delay jitters and communication overhead. Extensive literatures have been optimized for the end-to-end delay, of which [10–15] are more representative. The DT-DVTR algorithm [14] divides the system cycle of the satellite network system into N equal length time slots, and the topology of the network is deemed as fixed in each time slot. The end-to-end delay is used as the optimization target in each time slot, and multipaths are calculated for each pair of communication satellites to form the set of candidate paths. And then, a path is selected from the set to minimize the path switching between adjacent time slots. The DT-DVTR algorithm not only optimizes the end-to-end delay, but also takes into account the delay jitters caused by path switching. The drawback of this algorithm is that it cannot effectively solve the rerouting problem caused by link or connection switching.

The LZDR algorithm proposed in [12] models the polar orbit constellation to a Manhattan Street Network (MSN), and combines several adjacent nodes of MSN into one zone. The algorithm uses a two-stage routing method. The inter domain routing is calculated first and then the inner domain routing. The LZDR selects a representative node for each zone, and chooses the forwarding direction according to the hop metrics between the representative nodes. When a packet reaches the representative node, it is forwarded according to the precalculated forwarding direction. The delay-optimized path can be found this way. However, the MSN model can only be used in a polar orbit constellation, whose topology change is regular, and cannot be extended to inclined-orbit constellations. Therefore, it does not have extensive applicability.

The DRA algorithm [10, 11] and the MLSR algorithm [13] are also classical delay-optimized satellite network routing algorithms. As they have already been introduced in Chap. 1, we are not going to go into details here. The above-mentioned routing algorithms use delay or delay jitters as the optimization targets, and all achieve certain improvements. However, with the increasing diversification of satellite network business styles, the QoS requirements are getting more different now. Only optimizing one QoS requirement can no longer meet the needs of users. Especially when more than one QoS requirement is needed for the user, the above listed routing algorithms are even more powerless. In order to meet the QoS requirements of the users, joint optimization of multiple QoS parameters is needed in accordance with the requirements of the users. Thus, multiobjective optimization has been proved to be an NPC (Nondeterministic Polynomial Complete) problem. So, the heuristic algorithms and some classical multiobjective optimization methods are introduced to solve the multi-QoS-objective joint optimization routing problem of the satellite network.

Of the heuristic algorithms, the ant colony algorithm, the beehive algorithm, and the genetic algorithm are often used in the problem of portfolio optimization. The taboo search algorithm is derived from the local search algorithm. The heuristic algorithms have a wide range of applications. However, to the best of our knowledge, it is the first time to try various heuristic algorithms to solve the multiobjective QoS routing problem for a satellite network. Some classical multiobjective optimization algorithms, such as the Prior-Order algorithm and the PEC algorithm, are also used to solve the multiobjective QoS routing problem for satellite networks. They are also compared with the heuristic algorithms.

5.2 Satellite Network Heuristic QoS Routing Algorithm

The heuristic algorithm is a kind of algorithm constructed by intuition or experience. It gives a feasible solution to each instance of the combinatorial optimization problem under an acceptable cost (referring to the calculation of time, store space, etc.). However, the difference between the feasible solution and the optimal solution cannot be predicted beforehand [16]. Another definition of the heuristic algorithm is that, the heuristic algorithm is a technology of finding the best solution under an acceptable computing cost. It may not be able to guarantee the feasibility and optimality of the solution obtained, or even in most cases, cannot elaborate the difference between the solution obtained and the optimal solution [17].

Although the quality of the solution of heuristic algorithms may not be predictable, in some cases, especially in some real situations, the computing time of the optimal algorithm cannot be tolerated, or increases at an exponential rate with the increase of the instance scale. In this situation, only heuristic algorithms can be used to find a feasible solution to the problem in the instance [16]. Specifically, in solving the satellite network routing problem in order to satisfy the multiple QoS requirements of different users at the same time, while the traditional SPF routing algorithm fails to do so, a heuristic routing algorithm would be a viable method. In order to use heuristic algorithms in the routing calculation for the satellite network, we first introduce the satellite network routing model.

5.2.1 Satellite Network Routing Model

The system cycle of the *Tr* constellation is divided into several snapshots according to the movement cycle of the LEO/MEO multilayered constellation using the NSGRP strategy proposed in Chap. 3. The normal routing service in NSGRP provides service only for the business that requires delay guarantees. It adopts a distributed routing algorithm and is run on MEO satellites. The QoS routing service in NSGRP provides service for the users that have multiple QoS requirements. It adopts a source routing model and is run at the ground base station. In the beginning of each snapshot, the ground base station runs the QoS routing algorithm to provide QoS service for the users.

We model the whole satellite network (including the *Tr* constellation and the ground base station) to a directed weighted graph $G(V, E)$, in which $V = \{v_1, v_2, \ldots, v_m\}$ is the set of all the satellite nodes and ground base stations, and $E = \{e_1, e_2, \ldots, e_n\}$ is the set of all the ISLs, IOLS, and UDLs. Assume that i, j are the nodes of the network (i, j) is the link from node i to node j, s and d represent the source node and destination node of the path, respectively, $path(s, d)$ is a path from the source node s to the destination node d, $delay()$ is the link delay or path delay. A link delay is composed of the queuing delay and the propagation delay. They are defined respectively as follows.

$$delay_q(i,j) = np(i,j) \times \frac{al}{cp(i,j)}, \tag{5.1}$$

where $delay_q()$ is the queuing delay of the link; $np(i, j)$ is the number of packets in the buffer of link (i, j) currently; al is the average length of the packets; $cp(i, j)$ is the capacity of link (i, j).

In order to reduce the communication overhead, the LEO satellite does not report the link state to the ground base station in real-time in NSGRP within the snapshot. So $np(i, j)$ here is a prediction value according to the previous data. The prediction method is that

$$np(i,j) = \frac{1}{\Delta t} \times \int_{t_p}^{t_p+\Delta t} np(t) \times \frac{al}{cp(i,j)}\, \mathrm{d}t, \tag{5.2}$$

where $np(t)$ is the function of time and flow according to the characteristic of previous traffic, and t_p is the time needed to be predicted.

The propagation delay of the link is

$$delay_p(i,j) = \frac{ld(i,j)}{C_L}, \tag{5.3}$$

where $delay_p()$ is the propagation delay of the link; $ld(i, j)$ is the Euclidean distance from node i to node j; C_L is the speed of light.

From Eqs. (5.1) and (5.3), we have the link delay

$$delay(i,j) = delay_p(i,j) + delay_q(i,j) \tag{5.4}$$

Assume that $band(i, j)$ is the residual bandwidth of link(i, j), $loss(i, j)$ is the packet loss rate of link(i, j) (here the loss rate is a prediction value according to the previous loss rate), $jitter()$ is the delay jitter of the path, and $utility()$ is the link utility or maximum link utility of the path. In the multi-QoS-objective optimization routing scheme, the optimization objectives are usually as follows:

(1) Delay constraint:

$$delay(path(s,d)) \le D_r$$

(2) Bandwidth constraint:

$$\min_{(i,j)\in path(s,d)} (band(i,j)) \ge B_r$$

(3) Loss rate constraint:

$$1 - \prod_{(i,j)\in path(s,d)} (1 - loss(i,j)) \le L_r$$

(4) Jitter constraint:

$$jitter(path(s,d)) \le J_r$$

(5) Utility constraint:

$$utility(path(s,d)) \le U_r$$

The constraint parameters D_r, B_r, L_r, and J_r are predefined constants according to the user's requirements. If the QoS constraints required are more than one, it is a multiobjective optimization problem which cannot be solved by the traditional SPF algorithm. The following sections describe how to use heuristic algorithms to solve the multiobjective routing optimization problem in detail. Only three important QoS requirements: delay, utility, and packet loss rate, are selected as our optimization objectives, other QoS requirements can be added when needed.

5.2.2 Ant Colony QoS Routing Algorithm

1. Principle of ant colony algorithm

The Ant Colony Optimization (ACO), algorithm is a novel heuristic algorithm first proposed by Italian scholar M. Dorigo et al. in 1991. It has the characteristics of parallelism and robustness [18]. The algorithm has already been used in solving the Travel Salesman Problem, TSP, and other combinatorial optimization problems. It is a bionic algorithm that simulates the behavior of seeking paths of the ants in nature [19].

The scientists found after tracking the behavior of ants that an ant colony can always find a shortest path from the nest to a food source within a certain period of time [16]. This is because ants leave some volatile chemicals, called pheromones on the paths they use when foraging. These pheromones can be perceived by the following ants in the same colony. They have an impact on their behavior of finding the correct paths [20]. Experiments show that the path with higher density of pheromones will be selected in higher probability by the ants than the path with lower density of pheromones. As shown in Fig. 4.1, the ants depart from their nest and randomly find a path leading to a food source. After they find the food source, they will return to the nest along the paths they came from according to their previously released pheromones. In this procedure, the ants that found shorter paths will release more pheromone on the paths than the ants that found longer paths after a period of time. The shorter paths will hence attract more ants. Through this positive feedback mechanism, the density of pheromones on the shorter paths will become higher and higher. Eventually, almost all the ants will travel through the shortest path.

2. Implementation of ant colony QoS routing algorithm

According to the NSGRP strategy described in Chap. 3, the LEO/MEO satellite nodes need to do the following at the beginning of each snapshot.

(1) The LEO satellite probes the indicators of the four ISLs (there are only three ISLs if the LEO satellite located on both sides of the reverse seam) connected to it, and reports the LMR to its manager MEO satellite.
(2) The MEO satellite probes the indicators of the ISLs connected to it, and reports them together with the *LMR* of the LEO satellites in its group to its neighboring MEO satellites.

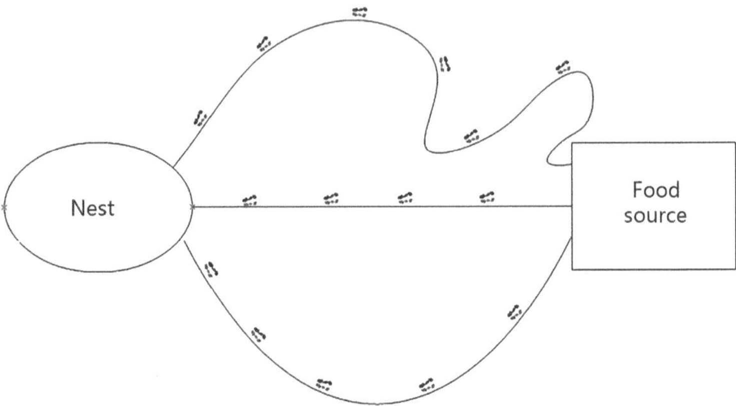

Fig. 5.1 Principle of ant colony algorithm

(3) When each MEO satellite has the network topology and the link state of the whole LEO/MEO constellation, it will broadcast to the ground base station in its coverage for the base station to compute the QoS routing.

Theoretically, heuristic algorithms can satisfy all the QoS requirements of the users. However, we only consider delay, utility, and the packet loss rate in this chapter, other QoS requirements can be added according to the application needs (Fig. 5.1).

The implementation of the ant colony QoS routing algorithm is as follows:

(1) Assume that the source node is s and the destination node is d, the starting node of the ant is i. The ant will choose the path according to the density of pheromones. Assume that the ant chooses link (i, j) with a probability $P(i, j)$, the density of link (i, j) is denoted as $\tau(i, j)$, the neighboring nodes of i are denoted as $N(i)$, so we have:

$$P(i,j) = \frac{\tau(i,j)}{\sum\limits_{j \in N(i)} \tau(i,j)}, \tag{5.5}$$

where $\tau(i,j) = [1/d(i,j)]^{\varepsilon_1} \times [r(i,j)/c(i,j)]^{\varepsilon_2} \times [1/l(i,j)]^{\varepsilon_3}$, $d(i, j)$ is the delay of link (i, j), $r(i, j)$ is the data rate of link (i, j), $c(i, j)$ is the capacity of link (i, j), $l(i, j)$ is the packet loss rate of link (i, j), and ε_1–ε_3 are the weight of delay, bandwidth and the packet loss rate.

(2) When an ant chooses a link that satisfies the requirements, it will release pheromones on it. Set ρ the attenuation factor between (0, 1), and α the routing impact factor, then we have

$$\tau(i,j) = (1-\rho)\tau(i,j) + \alpha \times \left[\frac{1}{d(i,j)}\right] \times \left[\frac{r(i,j)}{c(i,j)}\right] \times \left[\frac{1}{l(i,j)}\right] \tag{5.6}$$

(3) When an ant reaches the destination d, it will return by the way it came, and release pheromones on all the links it has passed through. Assume β is the path impact factor, for all the links (i, j) passed through we have

$$\tau(i,j) = (1-\rho)\tau(i,j) + \beta \times \sum_{i,j \in p(s,d)} \left[\frac{1}{d(i,j)}\right] \times \left[\frac{r(i,j)}{c(i,j)}\right] \times \left[\frac{1}{l(i,j)}\right], \tag{5.7}$$

where $p(s, d)$ is the set of paths from source s to destination d.

(4) When an ant returns to the source node, for the links (i, j) that are not selected we have

$$\tau(i,j) = (1-\rho)\tau(i,j) \tag{5.8}$$

As seen from the above-described process, the links with higher quality of service will have greater probability to be selected. The selected links have the bonus on pheromone as shown in Eq. (5.6). The selected paths have the bonus of pheromones according to the comprehensive quality of service as shown in Eq. (5.7). At the same time, the pheromones of the links that are not selected are just getting attenuated with the time as shown in Eq. (5.8). In this way, the paths that meet the QoS requirements will emerge from the feasible solutions after a period of positive feedback of the ants.

In order to implement the algorithm, we need to define the following variables: *through* is the array that stores the nodes passed by the ant; *needless* belongs to node s, and stores the nodes that are the neighbors of node s but do not meet the requirements; *delay* is the total delay of the links passed through; m is the number of cycles of the algorithm. The implementation of the algorithm is as follows:

Step 1: Initialize the pheromones of each link, and set each parameter properly. Create an artificial ant and place it at the source node.

Step 2: When an ant reaches node i, check the destination node d. If $i = d$, then a path from the source node s to the destination node d can be got from array *through*. Update the pheromone of the links in the path using Eq. (5.7). At the same time, update the pheromones of the links not in the path using Eq. (5.8), and exit the program. If m is larger than a certain threshold, the algorithm ends; otherwise turn to *step* 1. If $i \neq d$, choose the next node j in the probability defined in Eq. (5.5), and turn to *step* 3.

Step 3: If node $j \in$ *through*, refresh the pheromones of the link (i, j) using Eq. (5.8) and turn to step 2. If node $j \notin$ through, check if $i \in$ *needless*, if so, refresh the pheromones of (i, j) using Eq. (5.8) and turn to step 2. If the number of attempts of the ant to find the next hop j at node i exceeds k, then get the previous hop of node i from *through* and turn to step 2, otherwise turn to step 4.

Step 4: Check if the path selected meets the QoS requirements. If so, calculate *delay* and add j to through, replace the current node i with node j and turn to step 2, otherwise refresh the pheromones of link (i, j) using Eq. (5.8), add node j to the *needless* of node i, and turn to step 2.

3. The improvement of ant colony QoS routing algorithm

As the topology of the satellite network changes very quickly, and the lengths of the snapshots in a system cycle are quite different, the algorithm needs a short convergence time. In order to shorten the convergence time, the adjustment of parameters alone is not enough. After a careful observation of the above steps, we will find that step 2 is a key step in determining the convergence time of the algorithm. In step 2, the pheromone refreshment of the selected links directly determines the convergence of the algorithm. If the selected links are given too high an amount of pheromones, the algorithm may fall into a local optimum and may not get satisfactory results. Conversely, if the selected links are given too low an amount of pheromones, the algorithm will have a long convergence time. What we have to do is to find a way that not only does not make the algorithm fall into a local optimum, but also shortens the convergence time of the algorithm. The main idea of this approach is to give the current optimal path selected by the ant an additional amount of pheromones. This gives the current optimal path appropriate incentives, which allows the algorithm to converge faster. Compared with the method that accelerates the converging speed just by giving a high amount of pheromones to the selected links, this approach does not fall into a local optimum easily.

As mentioned above, step 2 will be modified according to the idea of the approach. In this step, an item to record the current optimum is added. If the current node is the destination node, compare the current path with the current optimum path. If the current path is equal to or better than the current optimum path, refresh this path with Eq. (5.7); otherwise refresh the current optimum path with Eq. (5.7), and refresh the current path with Eq. (5.9)

$$\lambda_1\tau(i,j) + \lambda_2(1-\rho)\tau(i,j) + \beta \times \left\{ (1-\rho)\tau(i,j) + \sum_{i,j\in p(s,d)} \left[\frac{1}{d(i,j)}\right] \times \left[\frac{r(i,j)}{c(i,j)}\right] \times \left[\frac{1}{l(i,j)}\right] \right\}, \tag{5.9}$$

where $\lambda_1 + \lambda_2 = 1$.

We select a snapshot to test the improved ant colony algorithm. Set $\lambda_1 = \lambda_2 = 0.5$, $\alpha = 3$, $\beta = 0.03$, $\rho = 0.03$, $\varepsilon_1 = 0.5$, $\varepsilon_2 = 0.25$, $\varepsilon_3 = 0.25$. The results are shown in Fig. 5.2.

Ant colony algorithm is a novel heuristic algorithm that combines distributed computing, positive feedback mechanism, and greedy search. It has a strong ability to search for feasible optimal solution, and can find the satisfied solution quickly. Therefore, as a QoS routing algorithm, it can form the routing table in a short time and is able to avoid premature convergence during the routing computation. These are all advantageous for the achievement of optimum solution that meets the user's

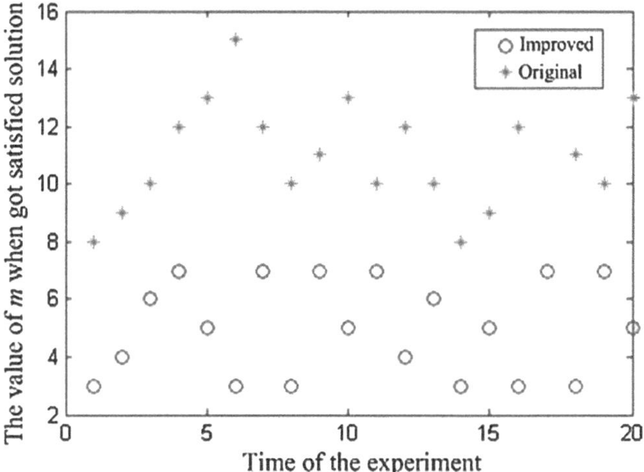

Fig. 5.2 Comparison between improved and original ACO

QoS requirements. However, there are some drawbacks in ant colony algorithm too. The convergence time depends on the parameter setting of the pheromone refreshment even using the improved algorithm. Besides, the algorithm is prone to be trapped after a period of time if the pheromone refreshment parameter is not set properly. The better solution cannot be found in this situation.

5.2.3 Taboo Search QoS Routing Algorithm

The taboo search algorithm is an extension of the local neighborhood search algorithm. It is a successful application of artificial intelligence in solving combinatorial optimization problems. The concept of taboo search was first proposed by Glover [21, 22] in 1986, and then it was implemented as an algorithm. The taboo search algorithm uses a taboo technology to prohibit repeat searching in several steps, in order to avoid falling into a local optimum. The so-called taboo technology is a taboo list that records the local optimums that have been reached and the process of reaching these optimums, which will help escaping local optimums [16].

The taboo search algorithm fully reflects the concentration and diffusion strategies. The concentration strategy is reflected in local search to seek a better solution from a point in its neighborhood to achieve a local optimum [16]. The diffusion strategy is reflected in the taboo list. In order to escape local optimums, the taboo list records the reached points and prohibits visiting them in certain steps. At the same time, visits to some points that have not been reached yet are encouraged to achieve a larger searching area.

Since the taboo search algorithm is an extension of the local neighborhood search algorithm, in order to understand the taboo search algorithm, we first introduce the local neighborhood search algorithm.

1. Local neighborhood search algorithm

First, the mathematical model of the optimization problem to be solved is given as follows:

$$\begin{aligned} &\min f(x) \\ &s.t. \quad g(x) \geq 0, \ x \in D, \end{aligned} \tag{5.10}$$

where $f(x)$ is the objective function; $g(x)$ is the constraint equation; D is the definition domain, here is a set of discrete points.

The algorithm to solve this optimization problem is as follows:

Step 1: Find an initial feasible solution x_{initial}, and set $x_{\text{best}} = x_{\text{initial}}$. Set $M = N(x_{\text{best}})$, here $N(x_{\text{best}})$ means the neighbors of x_{best}.

Step 2: If $M - \{x_{\text{best}}\} = \varnothing$, or the stop condition is satisfied, stop computing and output the computation results; otherwise, select a set P from M, and find the best solution x_{now} from P. If $f(x_{\text{now}}) < f(x_{\text{best}})$, then $x_{\text{best}} = x_{\text{now}}$, $M = N(x_{\text{best}})$; else $M = M - P$; turn to step 2.

The initial solution x_{initial} of the above algorithm can be randomly selected or got by experience. The stop condition is given by the algorithm designer according to the calculation time or the required results. The selection of P can affect the calculation speed and efficiency of the algorithm. A larger P means a large searching range and hence a large amount of calculation, whereas a smaller P means a smaller searching range and a small amount of calculation.

2. Taboo search algorithm

The taboo search algorithm is extended from the local search algorithm. It uses a taboo list to mark the reached local optimum solutions and the procedures of reaching local optimums. In further iterations, these solutions and procedures will be avoided in order to reach a global optimum.

The implementation of the taboo search algorithm is as follows:

Step 1: Set a taboo list $T = \varnothing$, and select an initial solution x_{initial}.

Step 2: If the stop condition is satisfied, stop computing and output the results; otherwise, find a candidate set $Can\_N(x_{\text{initial}})$ that is not on the taboo list from $N(x_{\text{initial}})$, from which a optimal solution x_{next} is selected, $x_{\text{initial}} = x_{\text{next}}$, update the list T and turn to step 2.

It is worth noting that the connectivity condition of the neighborhood is a necessary condition of the accessibility of global optimums. The connectivity condition is defined as follows [20].

Definition 5.1 A set C is connected under the neighborhood mapping N if and only if $\forall x,y \in C$, $\exists x = x_1, x_2, \ldots, x_l = y$, that $N(x_i) \cap \{x_{i+1}\} \neq \varnothing$, $i = 1, 2, \ldots, l - 1$.

Using the above basic idea of the taboo search algorithm, the taboo search used for the QoS routing can be constructed.

3. Implementation of the taboo QoS routing

In implementing the taboo search QoS routing algorithm, we still need to consider the following three QoS indicators.

(1) Link delay $delay(i, j) \le D_r$;
(2) Link utility $\theta_{ij} = r(i, j)/c(i, j) \le U_r$;
(3) Packet loss $loss(i, j) \le L_r$;

Therefore, the QoS routing algorithm in the *Tr* constellation can be described as follows: give three matrixes $\boldsymbol{D}_{ij}(n \times n)$, $\boldsymbol{U}_{ij}(n \times n)$, $\boldsymbol{L}_{ij}(n \times n)$, d_{ij} is an element of $\boldsymbol{D}_{ij}$, which represents the $delay(i, j)$ of link (i, j); u_{ij} is an element of $\boldsymbol{U}_{ij}$, which represents the θ_{ij} of link (i, j); l_{ij} is an element of $\boldsymbol{L}_{ij}$, which represents the $loss(i, j)$ of link (i, j). The goal is to find a path that has minimum delay, link utilization, and packet loss rate.

The objective function can be described as follows:

minimum $\text{delay}(s, d) = \sum d_{ij}$; $d_{ij} \in \boldsymbol{D}_{ij}$ $(i, j) \in \text{path}(s, d)$;

minimum $\text{utility}(s, d) = \max u_{ij}$; $u_{ij} \in \boldsymbol{U}_{ij}$ $(i, j) \in \text{path}(s, d)$;

minimum $\text{loss}(s, d) = 1 - \prod (1 - l_{ij})$; $l_{ij} \in \boldsymbol{L}_{ij}$ $(i, j) \in \text{path}(s, d)$;

The implementation of the taboo search QoS routing algorithm is as follows:

Step 1: Set the neighborhood of the taboo search 2-opt neighborhood, and prohibit backtracking of the found solutions in 3 steps. The initial solution x_{initial} is a feasible path from the source node to the destination node, $x_{\text{initial}} = x_{now}$.

Step 2: If $delay(i, j) \le D_r$, $\theta_{ij} = r(i, j)/c(i, j) \le U_r$, $loss(i, j) \le L_r$ in x_{now}, stop computing and output the result; otherwise, choose a x_{best} that has the optimal *fee* from the neighborhood set, $x_{now} = x_{\text{best}}$ and repeat step 2.

In the above implementation, the evaluation of the feasible solution *fee* is set as $fee = \alpha \times \text{delay}(s, d) + \beta \times r(s, d)/c(s, d) + \gamma \times 1/loss(s, d)$, where $\alpha + \beta + \gamma = 1$. If other QoS requirements are needed to be considered, the corresponding QoS indicators are added to the *fee*. This indicates that the taboo search algorithm is scalable. In practical applications, the weights α, β, and γ are adjusted according to the importance of each QoS indicator.

The implementation of the taboo search algorithm is simple, and can get a satisfactory solution quickly. This is very useful to quickly generate a QoS routing table. A quick generation of the routing table means a stable network state. The packet loss caused by link switching will be reduced accordingly. However, the performance of the taboo search algorithm mainly depends on the setting of taboo steps and the representation of the neighborhood, both of which rely on experience. The strong dependence between the quality of the path and the setting is a disadvantage of the taboo search algorithm.

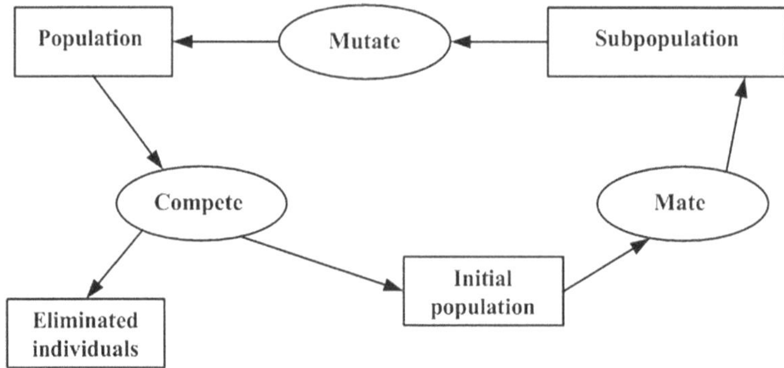

Fig. 5.3 Principle of the genetic algorithm

5.2.4 Genetic QoS Routing Algorithm

1. Principle of genetic algorithm

The genetic algorithm was first proposed in the early 1970s by Professor Holland. In 1975, Holland published his first book "Adaptation in Natural and Artificial Systems" which was the first book that systematically discussed the genetic algorithm.

Same as the ant colony algorithm, the genetic algorithm is also a bionic algorithm, which uses the "survival of the fittest" rules of the organisms in the evolutionary process. As shown in Fig. 5.3, after competition, some individuals in the population are eliminated for they cannot adapt to the environment, and the surviving individuals form the initial population. A subpopulation is got through mating of the individuals of the initial population. The subpopulation inherits the good genes of the initial population. The subpopulation becomes a new generation of population after mutation. Since the new population is formed through competition and mating of the older generation, it is more competitive than the older generation. Therefore, after several rounds of evolution, an excellent new population will be got.

The genetic algorithm solves and optimizes the problem mainly through three steps. First, encode the feasible solutions to the problem to be optimized. Each code of the solution is called a chromosome, and the elements of the chromosome are called genes. Second, construct a fitness function according to the optimization target. The fitness function can be the objective function of the problem. Determine which chromosomes are fit for survival and which are eliminated according to the fitness function of the chromosomes with certain probability. The surviving chromosomes form the initial population used for reproducing the next generation. Last, combine the chromosomes of the initial population. A new generation is generated through mating of the parent chromosomes. In the process of generating new solutions, some genes are changed with certain probability to enlarge the searching range of the feasible solution.

2. Implementation of the genetic algorithm

The genetic algorithm is implemented as follows:

Step 1: Give a coding rule for the problem to be solved, and find an initial population that has N chromosomes $POP(t)$, $t = 1$ is the generation.

Step 2: Compute the fitness function of each chromosome $pop_i(t)$ in $POP(t)$, $f_i = \text{fitness}(pop_i(t))$.

Step 3: If the stop condition is satisfied, the algorithm stops; otherwise, compute $p_i = f_i / \sum_{j=1}^{n} f_i$, $i = 1, 2, \ldots, N$, and choose some chromosomes from $POP(t)$ with the probability p_i to form a new initial population, New $POP(t + 1) = \{pop_j(t) | j = 1, 2, \ldots, N\}$, where New $POP(t + 1)$ can have the element in $POP(t)$ also.

Step 4: A cross population Cross$POP(t + 1)$ that contains N chromosomes is got through mating.

Step 5: Change several genes in some chromosomes with a low probability p, and form a population Mut$POP(t + 1)$; $t = t + 1$, a new population $POP(t) =$ - Mut$POP(t)$, and then turn to step 2.

3. Implementation of the genetic QoS routing algorithm

Same as the above two QoS algorithms, the related QoS metrics are still the following three:

(1) Link delay $delay(i, j) \leq D_r$;
(2) Link utility $\theta_{ij} = r(i, j)/c(i, j) \leq U_r$;
(3) Packet loss $loss(i, j) \leq L_r$;

Based on this, assume the source node of path i is s, the destination node of path i is d, path(s, d) is the set of links of path i, num(path(s, d)) is the number of elements of the set. The fitness function of path i can be defined as $f_i =$ fitness($pop_i(t)$) $= \alpha \times Di \mid \beta \times U_i + \gamma \times L_i$, where Di is the total delay of the path, $Di = \sum delay(x, y)$ $(x, y) \in$ path(s, d); U_i is the arithmetic average of the link utilities of path i, $U_i - (\sum r(x, y)/c(x, y))$/num(path($s$, d)) $(x, y) \in$ path(s, d); L_i is the sum of the reciprocal of the link packet loss in path i, $L_i = \sum 1/loss(x, y)$ $(x, y) \in$ path(s, d). α, β, and γ are weights of the QoS indicators, $\alpha + \beta + \gamma = 1$. In practical applications, the weights α,β, and γ are adjusted according to the importance of each QoS indicator. If other QoS requirements are needed to be considered, the corresponding QoS indicators are added to the fitness function. This indicates that the genetic QoS routing algorithm is also scalable.

Before further elaborating on the genetic QoS routing algorithm, we discuss the encoding rule, judging rule, mating rule, and mutation rule first.

Encoding rule Given an LEO group, a source satellite s, and a destination satellite d; assume there are m satellites in this group, and then there are maximum $m-2$ hops in a path. That is to say, in order to find a path from s to d, we need to consider the situations of 1, 2, …, $m-2$ hops in a path. Sort the feasible paths in the sequence of 1, 2, …, $m-2$ hops, and there will be $m-2$ kinds of paths with the hops 1, 2, …, $m-2$ separately. In the ith kind, the number of possible paths is the full array of the i hops. In each kind of paths, the possible paths are sorted using the 2-opt rule. Encode all the possible paths using the binary encoding method.

Judging rule For each possible path $(s_0, s_1, \ldots, s_{n-1}, s_n)$, $s_0 = s$, $s_n = n$, check if there exists a link from s_i to s_{i+1}, $i = 0, \ldots, n-1$. If all of them have a direct link, the path is a feasible path, otherwise it is not an available path.

Mating rule For the obtained binary codes, choose the $\lceil n/2 \rceil$ position to do simple cross mating.

Mutation rule Number the new generation obtained after simple cross mating, and choose 2 % of the chromosomes for mutation (In other words, the mutation probability is 2 %, the value can be adjusted according to the application.). For each chosen chromosome, invert a random position.

Based on the above rules, the genetic QoS routing algorithm is given as follows:

Step 1: Randomly choose an initial population $POP(t)$, $t = 1$ that contains N chromosomes, and encode the population with the encoding rule.

Step 2: For each chromosome $pop_i(t)$ in $POP(t)$, compute the fitness function f_i using the above-mentioned method. If the chromosome is not a feasible path according to the judging rule, its $f_i = 0$.

Step 3: If all the paths satisfy $D_i \le D_r$, $U_i \le U_r$, $L_i \le L_r$, where D_r, U_r, and L_r are all predefined constants, then stop computing and output the results, and choose the path with the highest fitness function; otherwise compute $p_i = f_i / \sum_{j=1}^{n} f_i$, $i = 1, 2, \ldots, N$, and select some chromosomes from $POP(t)$ with the probability distribution of p_i to form a new initial population $NewPOP(t) = \{pop_j(t) | j = 1, 2, \ldots, N\}$.

Step 4: Mate according to the mating rule and get the subpopulation $CrossPOP(t)$. The number of chromosomes in $CrossPOP(t)$ is the same as the $NewPOP(t)$.

Step 5: Mutate the chromosomes in $CrossPOP(t)$ using the mutation rule, and form a population $MutPOP(t)$, $t = t + 1$, the new population $POP(t) = MutPOP(t)$, and turn to step 2.

The main advantage of the genetic algorithm is the usage of the probability transfer rule, which can optimize for an uncertainty space. Compared with the general stochastic optimization method, the genetic algorithm does not search from a point along a line. It searches the whole solution space simultaneously, which can effectively avoid falling into local optimization points. The genetic algorithm does not have any special requirements for searching space (such as connectivity, convexity, etc.). The coding of the decision variable is the only operand of the algorithm, which is very suitable for the satellite network with time varying connectivity. However, the efficiency of the genetic algorithm depends a lot on the encoding method. The coding of the chromosomes may occupy the constrained onboard storage a lot. Moreover, the increase of the satellite nodes will make the length of the code increase exponentially, which results in poor scalability of the genetic algorithm. Nevertheless, in our *Tr* constellation, with the grouping strategy, the genetic QoS algorithm works pretty well as shown in the simulations.

5.2.5 Beehive QoS Routing Algorithm

1. Principle of beehive routing algorithm

The beehive algorithm is a novel swarm-intelligence heuristic algorithm proposed by Wedde for the first time in 2004 [23]. It is widely used in machine learning and organic operations after it was proposed. The principle of the beehive routing algorithm is similar to that of the ant colony routing algorithm [24]. Bees find nectar away from the hive. When they return from the nectar, they will exchange the information about the nectar on the dancing floor of the hive using the waggle dance. Different dance postures represent differences in the quality of and the distance to the nectar. The high quality nectars that are near the beehive will be mined more frequently.

In the beehive QoS routing algorithm of this chapter, the dancing floor is abstracted as the routing table in each node, and the waggle dance is implemented as the evaluation function. As observed in nature, most bees prefer to exploit the nectar near the hive, and only a few bees exploit the nectar far from the hive. Therefore, we adopt two kinds of bee agents in the routing algorithm: the long distance bee agent and the short distance bee agent marked as LDB and SDB respectively. SDB is in charge of collecting and evaluating the paths near the beehive, and LDB travels around the whole network to collect the routing information. The principle of the beehive QoS routing algorithm is summarized as follows:

(1) The whole network is organized into foraging regions and foraging zones. Each foraging region has a representative node. Each node records the foraging zones that can be reached by the SDB. Each nonrepresentative node sends its neighbor a copy of the SDB periodically to refresh the routing information.
(2) When a copy of a SDB reaches a new node, the routing information of this node is updated. The copy of the SDB is copied and forwarded to all the nodes except the source node. This process is repeated until the lifespan of the SDB expires. If an SDB reaches a node that has received the copy of this SDB before, this SDB will be removed.
(3) The LDB always has a longer lifespan cycle, a copy of which is sent by the representative node. The process is similar to (2).
(4) In order to make faster transmission of the routing information, the queuing priority of the bee agent is higher than the ordinary data packets. When the bee agent reaches or leaves a node, the routing information is refreshed, and the evaluation of the neighbor nodes can be calculated. Each node needs to keep the routing information for the reachable nodes in the foraging region and the representative nodes of the foraging zone. Besides, it also needs to store the affiliation relationship among the nodes in the foraging region.
(5) When a data packet reaches a node, its next hop is selected according to the probability calculated by the evaluation function. This is similar to the ant

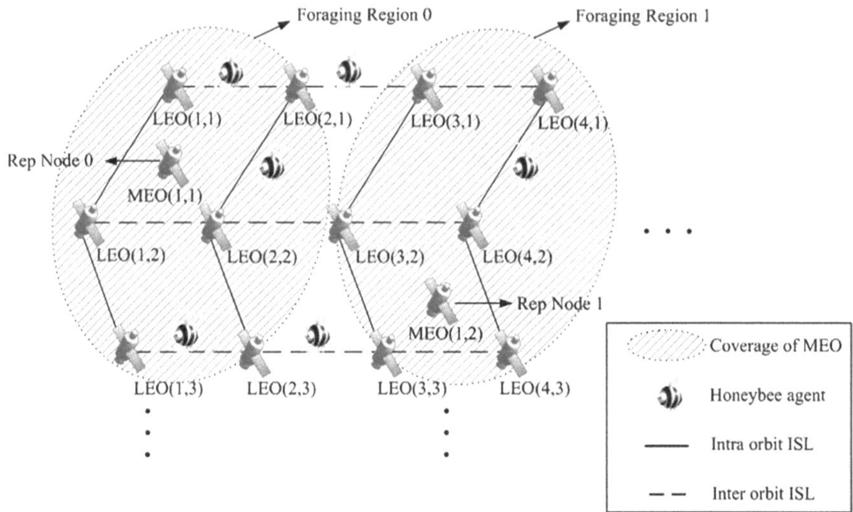

Fig. 5.4 Flooding structure of the beehive algorithm in the *Tr* constellation

colony algorithm, and is good for the balancing of the traffic load in the network.

2. Design and implementation of beehive QoS routing algorithm

Since the LEO/MEO double-layered constellation is the core structure of the *Tr* constellation, in the design of the beehive QoS routing algorithm for the *Tr* constellation, the LEO group in each snapshot is organized as a foraging region, and the MEO group manager is designated as the representative node of this foraging region. A foraging zone of a node contains all the nodes within 3 hops from it. The beehive flooding structure of the *Tr* constellation is shown in Fig. 5.4.

The LEO and MEO nodes shown in Fig. 5.4 are all virtual nodes of the ground topology graph. Since the QoS algorithms are carried by the terrestrial gateway, the bee agents don't travel around the LEO/MEO constellation. The beehive QoS algorithm runs at the terrestrial gateway.

As shown in Fig. 5.4, each virtual satellite node i keeps three kinds of routing tables: Intra Foraging Zone (IFZ) routing table, Inter Foraging Region (IFR) routing table and Foraging Region Membership (FRM) routing table. The IFZ routing table R_i is a $|N(i)| \times |D(i)|$ matrix, where $N(i)$ is the set of neighbor nodes of virtual satellite node i, and $D(i)$ is the set of destination nodes in the foraging zone. Each element Q_{jk} of R_i is a vector $(q_{jk}(1), \ldots, q_{jk}(l), \ldots, q_{jk}(m))$, where m is the number of user's QoS requirements, and $q_{jk}(l)$ is the lth evaluation function of the path from neighbor node j to the destination node k. Here, in order to compare with ant colony routing algorithm, we set $m = 3$. Table 5.1 shows an example of R_i. The routing table of IFR is similar to the IFZ routing table. $D(i)$ of IFZ is the set of MEO satellite nodes, and FRM contains the membership of LEO satellites and their managers.

Table 5.1 *IFZ* routing table

R_i	$D_1(i)$	$D_2(i)$	...	$D_d(i)$
$N_1(i)$	Q_{11}	Q_{12}	...	Q_{1d}
.	.	.	.	.
.	.	.	.	.
.	.	.	.	.
$N_n(i)$	Q_{n1}	Q_{n2}	...	Q_{nd}

As with the ant colony QoS routing algorithm, delay, link utility, and packet loss rate are also selected as the QoS optimization targets of the beehive algorithm. $q_{jk}(1)$ is the delay evaluation function of the path from neighbor node j to the destination node k, $q_{jk}(2)$ is the utility evaluation function of the path from neighbor node j to the destination node k, and $q_{jk}(3)$ is the packet loss evaluation function of the path from neighbor node j to the destination node k. The weight p_{jk} is similar to the weights in the ant colony routing algorithm, which is defined as:

$$p_{jk} = q_{jk}(1)^{e_1} \times q_{jk}(2)^{e_2} \times q_{jk}(3)^{e_3} / \sum\nolimits_{l=1}^{n} q_{lk}(1)^{e_1} \times q_{lk}(2)^{e_2} \times q_{lk}(3)^{e_3}, \quad (5.11)$$

where e_1, e_2, $e_3 < 0$ are the weights of delay, link utility, and packet loss, respectively. Since the membership in each snapshot is calculated in Chap. 3, FRM can be deemed as known. The pseudo code of the beehive QoS routing algorithm is given in Table 5.2.

Through this algorithm, each virtual node establishes its IFZ and IFR. When a virtual node receives a data forwarding requirement, it first checks if the IFZ destination node is in the foraging zone. If it is, choose the next hop according to the probability calculated by (5.11); otherwise find r in FRM and choose the next hop according to the probability calculated by (5.11) in IFR.

5.3 Satellite Network Multi QoS Optimization Routing Algorithm

Besides the heuristic algorithms, there are also some multiobjective optimization methods. These methods first define the multiobjective function $f(x_1, \ldots, x_n)$, and then define the concept of the solution to multiobjective optimization problems. The concepts of efficient solution, weak efficient solution, proper efficient solution, and strong efficient solution are also proposed, based on which the multiobjective optimization algorithms, such as the prior order algorithm and the PEC algorithm, are proposed in the subsection. We utilize the above two algorithms to solve the multi-QoS routing optimization problem.

In the multi-QoS optimization routing algorithm of the satellite network, the *Tr* constellation and terrestrial gateways are modeled as a directed weighted graph G (V, E), $V = \{v_1, v_2, \ldots, v_m\}$ is the set of nodes in the satellite network,

Table 5.2 Beehive QoS routing algorithm

Beehive QoS algorithm:

```
t:=current time, t_end :=end time of the algorithm
//Short_hop_limit:=LEO number in each group, Long_hop_limit:=
number+max(LEO number in each group)
// i=current node, s=source node, d=destination node
// p=predecessor node of i
// r=group leader MEO of s
// q(1)=delay evaluation estimated by bee agent from s to p
// q(2)=link utilization evaluation estimated by bee agent from s to p
// q(3)=package loss ratio evaluation estimated by bee agent from s to p
// Δt :=Bee_Generation_Interval, h_i:=hop limit for bees sent from i
// d_ip:=delay of link (i, p), u_ip:=utilization of link (i, p), l_ip
of link (i, p)
for each Node
  while (t<t   )

      if (t mod  Δt=0) // send a bee agent
         if (i is a LEO satellite)
            h_i :=Short_hop_limit, // b_i is a short distance agent
         else
            h_i :=Longt_hop_limit, // b_i is a long distance agent
         endif
            Send the bee agent b_i to all neighors in N(i)
      Endif
   for each agent b_s arrived at i from p // handle the received bee agent
      if (s=i or b_s already visited node i or h_s=hop limit)
           Remove the bee agent b_s from the system
      elseif (h_s=Short_hop_limit) //s is inside the foraging zone of node i
           q(1):= q(1)+ d_ip, q(2):= q(2)+ l_ip, q(3):= q(3)+ u_ip
           update IFZ entry q_ps(1)= q(1), q_ps(2)= q(2), q_ps(3)= q(3)
           q(1):= Σ_{l∈N(i)} q_ls(1)×g_ls, q(2):= Σ_{l∈N(i)} q_ls(2)×g_ls, q(3):= Σ_{l∈N(i)} q_ls(3)×g_ls

      else //s is a MEO node
           q(1):= q(1)+ d_ip, q(2):= q(2)+ l_ip, q(3):= q(3)+ u_ip
           update IFR entry q_ps(1)= q(1), q_ps(2)= q(2), q_ps(3)= q(3)
           q(1):= Σ_{l∈N(i)} q_lr(1)×g_lr, q(2):= Σ_{l∈N(i)} q_lr(2)×g_lr, q(3):= Σ_{l∈N(i)} q_lr(3)×g_lr

      endif
           Send bee agent b_s to all neighbors of i except p
   endfor
  endwhile
endfor
```

and $E = \{e_1, e_2, \ldots, e_n\}$ is the set of UDLs, ISLs, and IOLs. The QoS requirements, such as delay, residual bandwidth, and packet loss rate, are weights of the links.

Assume that $\boldsymbol{D}_{m \times m} = [d_{ij}]$ is the delay matrix of the satellite network system, d_{ij} is the total delay of link (i, j), that is the sum of propagation delay and queuing delay of the link; $\boldsymbol{B}_{m \times m} = [b_{ij}]$ is the bandwidth matrix of the system, b_{ij} is the residual bandwidth of link (i, j); $\boldsymbol{L}_{m \times m} = [l_{ij}]$ is the packet loss matrix of the system, l_{ij} is the packet loss of link (i, j).

When a terrestrial gateway receives a user packet, its source node s and destination node d will first be extracted. Assume that $\boldsymbol{X}_{m \times m} = [x_{ij}]$ is the path matrix, $x_{ij} = \{0, 1\}$, $\Re = \{ X_{m \times m} \mid x_{ij} = 1$,if link (i, j) is within the path from s to d; else $x_{ij} = 0\}$, $P = \{p_1, p_2, \ldots, p_w\}$ is the set of all the feasible paths of the OD gateways (s, d). For each of the OD gateways (s, d), the multi-QoS optimization routing algorithm can be described as:

$$\min\,(D_{m\times m}\cdot X_{m\times m},\ -B_{m\times m}\cdot X_{m\times m},\ L_{m\times m}\cdot X_{m\times m})^T$$
$$s.t.\ \begin{cases} D_{m\times m}\cdot X_{m\times m}\leq\alpha \\ B_{m\times m}\cdot X_{m\times m}\geq\beta \\ L_{m\times m}\cdot X_{m\times m}\leq\gamma, \\ x_{ij}\cdot b_{ij}\geq\delta,\quad (i,j)\in p_k,\ k\in[1,w] \\ X_{m\times m}\in\Re \end{cases} \tag{5.12}$$

where the parameters α, β, γ are set according to the application, and δ is the bandwidth required by the users. For the above multiobjective optimization problem described in (5.12), we transform its feasible solutions to a finite set, and use the prior-order and PEC algorithm [25] to solve it.

5.3.1 Prior-Order QoS Routing Algorithm

From the definition of the problem, we know that given an OD pair (s, d), if P is known, $\Re$ can be determined. P can be got by a constraint flooding algorithm. Before the introduction of the algorithm, we first define the variables used in the algorithm.

Definition 5.2 *Path* is a variable that stores the passed paths in the procedure of flooding.

Definition 5.3 *SumD*(*Path*) is an operation that sums the link delay in *Path.*

Definition 5.4 *SumB*(*Path*) is an operation that sums the link residual bandwidth in *Path.*

Definition 5.5 *SumL*(*Path*) is an operation that sums the link packet loss rate in *Path.*

Definition 5.6 $\boldsymbol{R}_{m \times m} = [r_{ij}]$ is a connection matrix. If there exists a direct connection between node i and node j, then $r_{ij} = 1$; otherwise $r_{ij} = 0$.

The pseudo code of the constraint flooding algorithm is shown in Table 5.3.

All the paths from s to d that satisfy the constraint of Eq. (5.12) can be got by $s \rightarrow$*path*, Flooding(s, *Path*). When all the possible paths are found, the prior-order algorithm can be used to solve the problem of Eq. (5.12).

Prior-order routing algorithm

The model of the prior-order algorithm is as follows:

The feasible domain $F = \{X_1, X_2, \ldots, X_u\}$

(VP) $\min_{X\in F} f(X) = (f_1(X), f_2(X), \ldots, f_q(X))^T$, $q > 1$.

Table 5.3 Pseudo code of constraint flooding algorithm

Constraint flooding algorithm:

```
Flooding (node i, Path)
{
    if (i==d)
    {
            break;
      print Path;
      }
    else
      {
        for (j=0; j<m; j++)
          {
              if (r_ij==1& SumD(Path) ≤ α & SumB(Path)
                      ≥ β SumL(Path) ≤ γ &b_ij ≥ δ )
                {
                      j→Path;
                      Flooding (j, Path);
                      }
                }
        }
}
```

Definition 5.7 $Q = \{1, 2, \ldots, q\}$, $U = \{1, 2, \ldots, u\}$, $\forall X_i, X_j \in F$, define variable

$$c_{ijl} = \begin{cases} 1, & f_l(X_i) < f_l(X_j) \\ 0.5, & f_l(X_i) = f_l(X_j) \\ 0, & f_l(X_i) > f_l(X_j)\ or\ i = j \end{cases},$$

$$c_{ij} = \sum_{l \in Q} c_{ijl}, \quad i, j \in U$$

Variable c_{ij} is the prior-order of the ith feasible solution X_i compared with the jth feasible solution X_j on all the objective functions $f_i(X)$. Accordingly, variable c_{ij} is also the inferior-order of the jth feasible solution X_j compared with the ith feasible solution X_i on all the objective functions $f_i(X)$. $A_i = \sum_{j \in U} c_{ij}$ is called the sum of the prior-order of the ith feasible solution X_i compared with all the feasible solutions on all the objective functions $f_i(X)$. Accordingly, $Z_j = \sum_{i \in U} c_{ij}$ is called the sum of the inferior-order of the jth feasible solution X_j compared with all the feasible solutions on all the objective functions $f_i(X)$.

According to the definition of A_i and Z_i, A_i represents the comprehensive quality of X_i on all the objective functions. That is to say, a larger X_i means a closer distance between the objective function $f_i(X_i)$ and the optimum.

In the multiobjective optimization QoS routing problem defined in Eq. (5.12) , $\Re$ can be got by *Flooding*(s, *Path*). $X_1, X_2, \ldots, X_u$ are path matrixes, each of which is a path calculated by *Flooding*(). That is, $u = w$, $X_1, X_2, \ldots, X_u \in \Re$.

Table 5.4 Calculation of the prior-order

X_j / X_i	X_1	X_2	…	X_j	…	X_u	A_i
X_1	0	c_{12}	…	c_{1j}	…	c_{1u}	A_1
X_2	c_{21}	0	…	c_{2j}	…	c_{2u}	A_2
⋮	⋮	⋮		⋮		⋮	⋮
X_i	c_{i1}	c_{i2}	…	c_{ij}	…	c_{iu}	A_i
⋮	⋮	⋮		⋮		⋮	⋮
X_u	c_{u1}	c_{u2}	…	c_{uj}	…	0	A_u
Z_j	Z_1	Z_2	…	Z_j	…	Z_u	

$f_1(X) = D.X, f_2(X) = -B.X, f_3(X) = L.X, q = 3$. A_i and Z_j can be calculated by Table 5.4.

Sort A_i when it is calculated. Choose the path that has the best A_i as the routing result. The path that has the suboptimal A_i is chosen as a backup routing. When a GEO satellite finds some satellites or links fail, it will recalculate the snapshots. The multi-QoS routing will be restarted accordingly.

5.3.2 PEC Routing Algorithm

The PEC algorithm is the abbreviation of Proper Equal Constraint algorithm. The main idea of PEC is to add $p-1$ equality constraints ($f_i(x) = a_i$ ($i = 1, 2, \ldots, p-1$),a_i is constant) on the (VP) constraint set R, and optimize $f_p(x)$ on the new constraint set $R_a = \{x \in R |\ f_i(x) = a_i,\ i = 1, 2, \ldots, p-1\}$. That is, to solve

$$(\text{PECSO}) \min f_p(x), x \in R_a \tag{5.13}$$

When a_i satisfies certain conditions, the optimal solution to PECSO is the efficient solution to (VP) [25].

When the prior-order algorithm is applied to the multi-QoS objective optimization routing of a satellite network, we model the p QoS requirements optimization routing as follows:

$$\min\left(f_1(x), f_2(x), \ldots, f_p(x)\right)^T$$

$$s.t. \begin{cases} g_i(x) \le 0 & (i = 1, 2, \ldots, m) \\ h_i(x) = 0 & (j = 1, 2, \ldots, l) \end{cases} \tag{5.14}$$

If

$$f(x) = (f_1(x), f_2(x), \ldots, f_p(x))^T$$

$$g(x) = (g_1(x), g_2(x), \ldots, g_m(x))^T$$

$$h(x) = (h_1(x), h_2(x), \ldots, h_l(x))^T$$

$$R = \{x \in E^{\mathrm{n}} | g(x) \leq 0, h(x) = 0\},$$

then model (5.14) can be abbreviated as

$$\text{(VP)} \min_{x \in R} f(x).$$

If $p = 1$, then (VP) is a simple single QoS objective optimization; otherwise if $p > 1$ (VP) is a multi-QoS requirements optimization. The essence of using PEC to solve the multi-QoS requirements routing is to gradually eliminate the solutions which cannot be efficient solutions. It is a direct method to find the set of efficient solutions.

According to the routing model (5.12) and the standard form of the multi-QoS objective optimization (5.14), $X_{\mathrm{m \times m}}$ is the decision variable of the algorithm, and equals to x in (5.14), $p = 3$. The constraint $x_{ij} \cdot b_{ij} \geq \delta$ is ignored in the calculation, and will be considered after the efficient solutions to (5.12) are found.

Therefore, we have

$$f_1(X_{\mathrm{m \times m}}) = D_{\mathrm{m \times m}} X_{\mathrm{m \times m}},$$

$$f_2(X_{\mathrm{m \times m}}) = -B_{\mathrm{m \times m}} X_{\mathrm{m \times m}},$$

$$f_3(X_{\mathrm{m \times m}}) = L_{\mathrm{m \times m}} X_{\mathrm{m \times m}},$$

$$g_1(X_{\mathrm{m \times m}}) = D_{\mathrm{m \times m}} X_{\mathrm{m \times m}} - \alpha$$

$$g_2(X_{\mathrm{m \times m}}) = -B_{\mathrm{m \times m}} X_{\mathrm{m \times m}} + \beta$$

$$g_3(X_{\mathrm{m \times m}}) = L_{\mathrm{m \times m}} X_{\mathrm{m \times m}} - \gamma$$

The steps of the algorithm are as follows:

Step 1: Transform (VP) (5.12) to (PECSO) (5.13). Assume $a = (a_1, a_2)$, in order to make sure that R_a is not null, set $\overline{R} = \{a \in E^2 | R_a \neq \emptyset\}$.

Step 2: Determine $\lambda(a) = \inf \{f_p(x) | x \in R_a, a \in \overline{R}\}$ in $\overline{R}$, and find solution $\widetilde{x}(a)$ that $f_p(\widetilde{x}(a)) = \lambda(a)$. If $\widetilde{x}$ satisfies the following conditions, it is an optimal solution to (PECSO) (5.13):

(1) $\widetilde{x} \in R_a$
(2) $f_p(\widetilde{x}) = \lambda(a)$
(3) $\lambda(a) > -\infty$

Step 3: Determine set $B = \{a| \ \lambda(a) > -\infty, \ \text{for} \ -\tilde{x} \in R_a, a \in \overline{R}, \exists f_p(\tilde{x}(a)) = \lambda(a)\}$. Obviously, if $\tilde{x}(a)$ is an optimal solution to (PECSO) (5.13), then $a \in B$; otherwise, for each $a \in B$, $\exists \tilde{x}(a) \in R_a$ is an optimal solution to (PECSO) (5.13).

Step 4: Assume $a^{''} \in B$, if $\exists$ a neighborhood $N(a^{''},\theta) = \{a \in E^2 | \|a - a^{''}\| < \theta, \theta > 0\}$ satisfies:

(1) If $\lambda(a) \leq \lambda(a^{''}), a \in \overline{R} \cap N(a^{''}, \theta)$, then $a \geq a^{''}$.

(2) $a^{''}$ fits locally, if for each $a \in \overline{R} \cap N(a^{''}, \theta)$ that $\lambda(a) \leq \lambda(a^{''})$, then $a \geq a^{''}$; then remove all the vectors from B that are not locally fit. The resultant set is marked as B^*.

Step 5: When $a \in B^*$, solve the optimal solution $\tilde{x}(a)$ to (PECSO) (5.13). Here $\tilde{x}(a)$ is the efficient solution to (VP) (5.12) when there is no constraint $x_{ij} \cdot b_{ij} \geq \delta$.

Notice that the efficient solution of step 5 is not the final solution for the user. The efficient solutions that do not satisfy the constraint $x_{ij} \cdot b_{ij} \geq \delta$ are removed. The resultant set F^* is the set of the solutions that satisfy all the QoS requirements of the users.

5.4 Simulations and Results

In order to investigate the performance of the heuristic QoS routing algorithm and the multi-QoS objective routing algorithm as the routing part of the NSGRP used for the *Tr* constellation, we design the following simulations. Among the multi-QoS routing algorithms mentioned in this chapter, the ant colony routing algorithm and the beehive routing algorithm belong to the swarm intelligence algorithms. To simplify the comparison, we only need to compare the ant colony routing algorithm (representing the swarm intelligence algorithm) and the other two heuristic routing algorithms, and then compare the ant colony routing algorithm and the beehive routing algorithm. And in the end, we choose a representative heuristic routing algorithm and compare it with the multiobjective routing algorithms.

All of the satellite network QoS routing algorithms mentioned in this chapter will be compared with the SPF routing algorithm. Currently, most satellite network routing protocols use the Distance Vector, DV, or Link State, LS, routing algorithms [8, 26–28], which are the same as the terrestrial network routing protocols. The RIP and OSPF algorithms are two representative algorithms of these two kinds of algorithms, and belong to the SPF algorithm. The SPF algorithm used in the following simulations is described in [8].

The constellation model used in the simulations is the *Tr* constellation, which is described in Table 2.3. We will compare the performance of the ant colony, taboo search, genetic, beehive, prior-order, and PEC routing algorithms. First, the *Tr* constellation model is established. The sub-satellite points of some satellites in the first snapshot of the NSGRP are shown in Fig. 5.5.

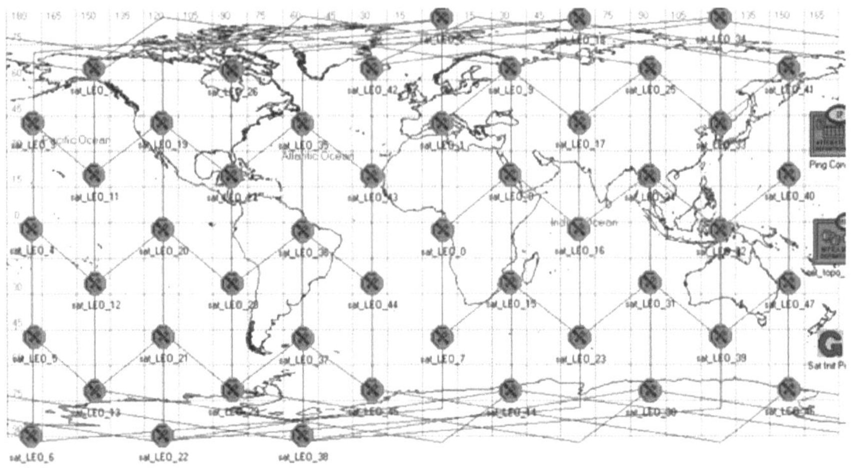

Fig. 5.5 Sub-satellite points of part of the *Tr* constellation

Assume that the demand of the network is distributed evenly. The evenly distributed 50 pairs of terrestrial users send the calling requirements in the same probability. The service time of the callings follows a negative exponential distribution with the average value 5 min. The LEO satellites provide access service for the ground users. The ISLs, IOLs and UDLs are all full duplex links. The bandwidth of ISLs and IOLs is 155 Mbps, and the bandwidth of UDLs is 5 Mbps. The core structure of the *Tr* constellation is the LEO/MEO double-layered satellite network, whose system cycle is 12 h. The caches of the LEO, MEO, and GEO satellites are 5 MB, 10 MB, and 20 MB, respectively. The parameters of the routing model are $D_r = 150$ ms, U_r=70 %, L_r=0.3 %. The QoS requirements of the routing model are randomly generated under the constraint of the parameters.

First, we compare the convergence times of the ant colony routing algorithm, the taboo search routing algorithm, and the genetic routing algorithm. As shown in Fig. 5.6, the convergence time of the taboo search routing algorithm is the shortest. This is because the taboo search derives from the local search, which has a short convergence time, and its calculation procedure is relatively simple compared with the ant colony routing algorithm and the genetic routing algorithm. The convergence time of the genetic routing algorithm is the longest. This is because the genetic algorithm is a kind of global search algorithm, and its coding rule and mating rule are relatively complex. Notice that the convergence time shown in Fig. 5.6 can only show the convergence time under certain parameters and certain implementation methods, all of the three comparison algorithms can reduce the convergence time through adjusting the algorithm parameters or modifying the implementation procedure.

The link congestions of the SPF, ant colony, taboo search, and genetic routing algorithms are shown in Fig. 5.7. As shown in Fig. 5.7, the performances of the three heuristic routing algorithms are better than that of the SPF algorithm. This is

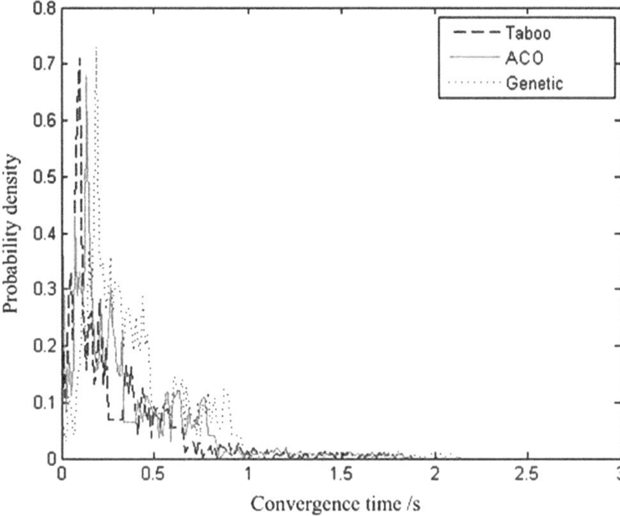

Fig. 5.6 Convergence time of heuristic algorithms

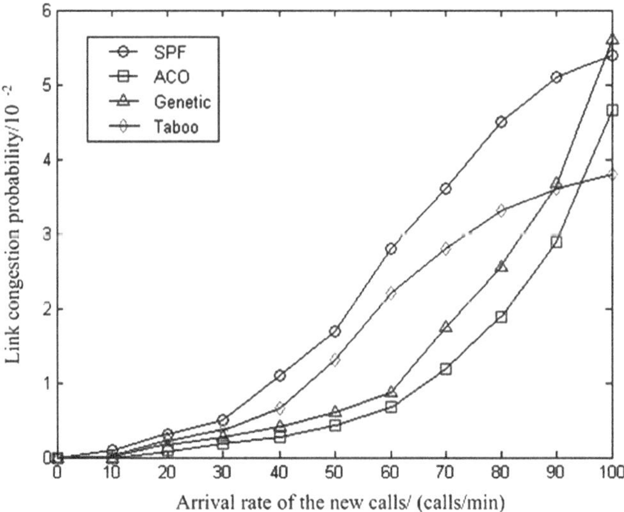

Fig. 5.7 Comparison of link congestion probability

because heuristic algorithms choose the path in probability according to the user's QoS requirements, which efficiently balance the load of the network, and reduce the congestion. However, the SPF algorithm can only satisfy the delay requirement of the users, which cause a low quality of service. Moreover, the concentration of all business on the shortest delay path may cause link congestion easily.

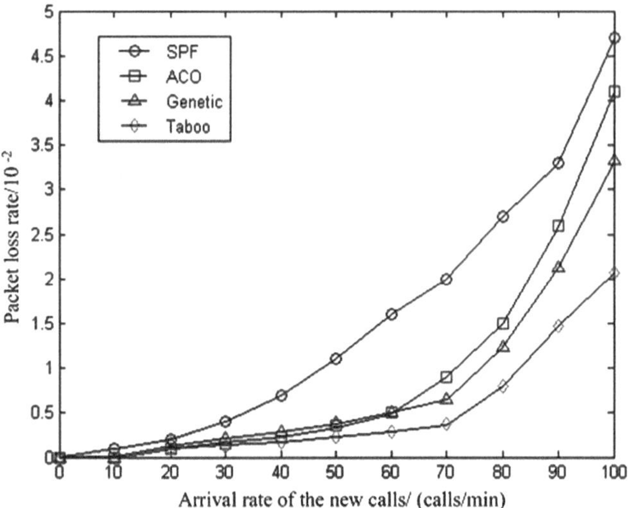

Fig. 5.8 Comparison of packet loss rate

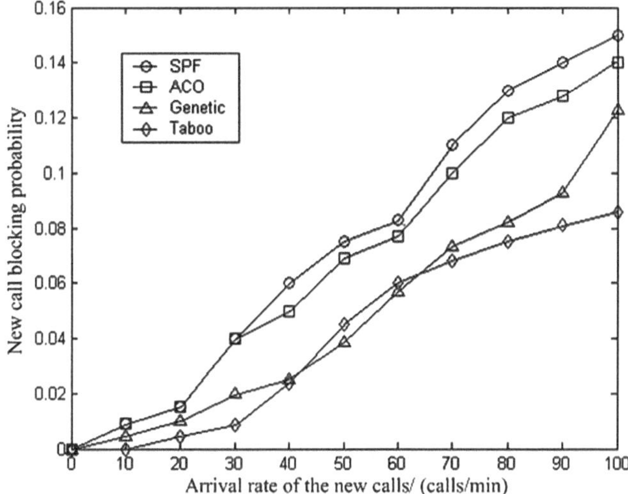

Fig. 5.9 Comparison of new call blocking probability

The packet losses of the algorithms are shown in Fig. 5.8. Since the heuristic algorithms consider the history of the link packet loss, they select the links that are more reliable. Hence, they perform better than the SPF algorithm overall. In addition, heuristic algorithms can avoid congestion to some extent, which is another reason for low link packet loss.

The new call blocking probabilities of the routing algorithms are shown in Fig. 5.9, while the call switching blocking probabilities of the routing algorithms

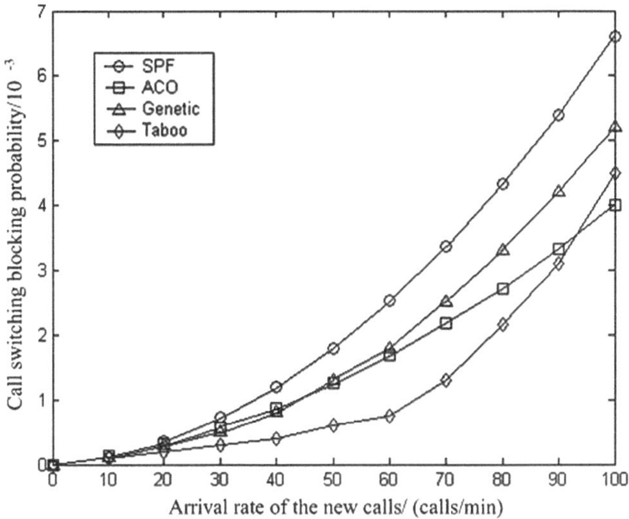

Fig. 5.10 Comparison of call switching blocking probability

are shown in Fig. 5.10. The heuristic routing algorithms perform better than the SPF routing algorithm under the same new call arrival rate. This is because heuristic algorithm can efficiently balance the load of the network. Most links hence have enough residual bandwidth to accept the new call and call switching requirements.

In order to evaluate the quality of the various types of QoS routing algorithms, we introduce an evaluation index *EV*:

$$EV = t/1000 + 100(P_c + P_l), \tag{5.15}$$

where t is the end-to-end delay/ms; P_c is the link congestion probability; P_l is the packet loss. *EV* is a comprehensive evaluation parameter, which reflects the performance of an algorithm. All the parameters in (5.15) are considered in the heuristic algorithms.

The results in Fig. 5.11 show that the taboo search QoS routing algorithm performs best under the high traffic load. The SPF algorithm has a performance quality similar to that of the heuristic algorithms. However, unlike the heuristic algorithms, the SPF algorithm can not provide QoS guarantee under a high traffic load.

The comparison of link congestion between the ant colony algorithm and the beehive algorithm is shown in Fig. 5.12, while the comparisons of packet loss, new calling blocking and call switching blocking are shown in Figs. 5.13, 5.14 and 5.15. The swarm intelligence algorithms perform better than the SPF algorithm overall. The reason is the same as the heuristic algorithms. The ant colony routing algorithm performs better than the beehive routing algorithm in link congestion,

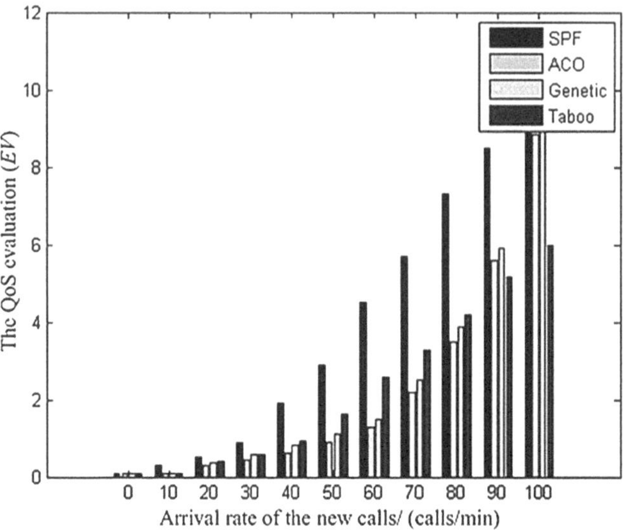

Fig. 5.11 Comparison of the QoS evaluation

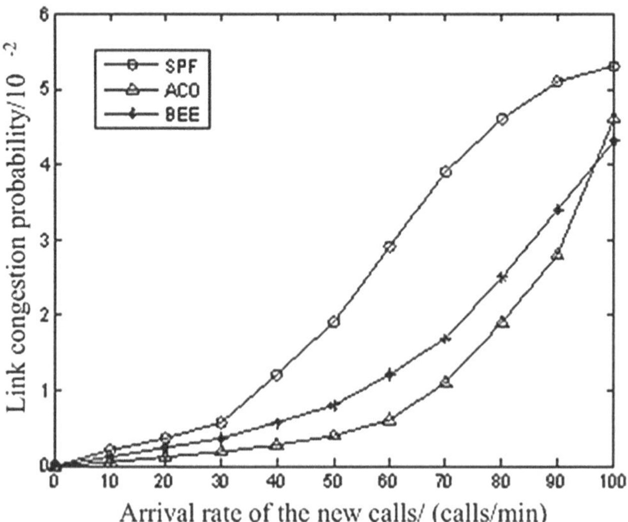

Fig. 5.12 Comparison of the link congestion probability

packet loss and new call blocking. However, the performance of the ant colony algorithm deteriorates dramatically when the load is very high. This shows that the positive feed back mechanism of the pheromone of the ant colony algorithm can strengthen the optimal paths more effectively than the beehive algorithm, and can balance the load. However, the complex pheromone positive feed back mechanism

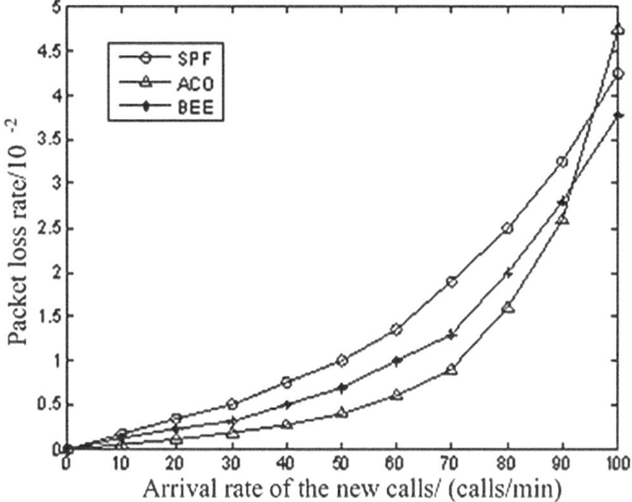

Fig. 5.13 Comparison of the packet loss rate

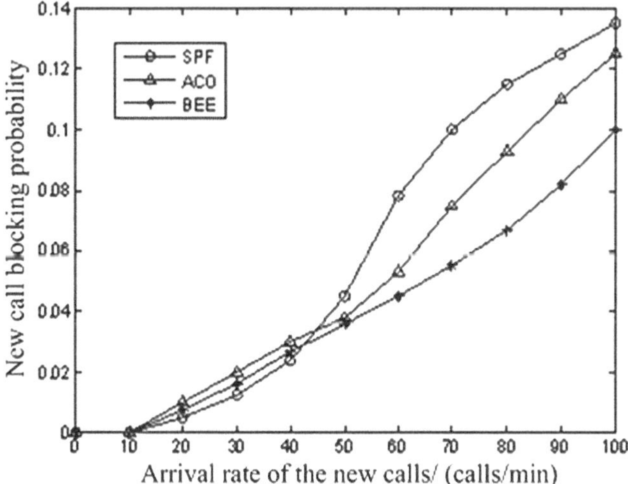

Fig. 5.14 Comparison of the new call blocking probability of swarm intelligence algorithms

may not have enough response time when the traffic is heavy, which causes dramatic performance deterioration. The ant colony algorithm performs better than the beehive algorithm during the whole procedure in call switching blocking. The reason is further studied below.

Figures 5.16, 5.17, 5.18, and 5.19 show the performance comparison of the ant colony algorithm on behalf of the heuristic algorithms with two multiobjective

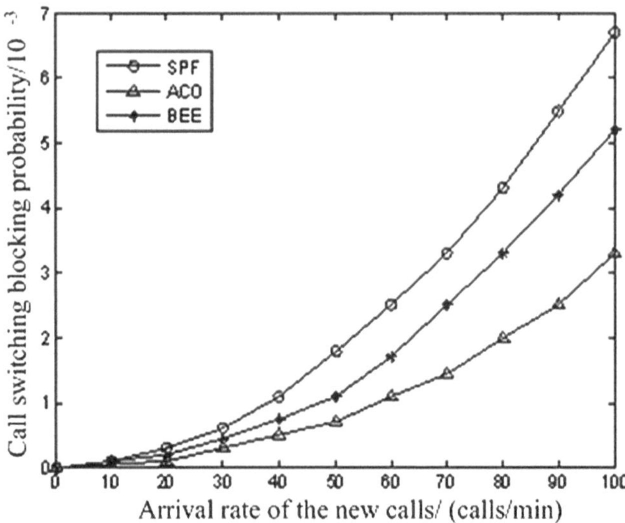

Fig. 5.15 Comparison of the call switching blocking probability of swarm intelligence algorithms

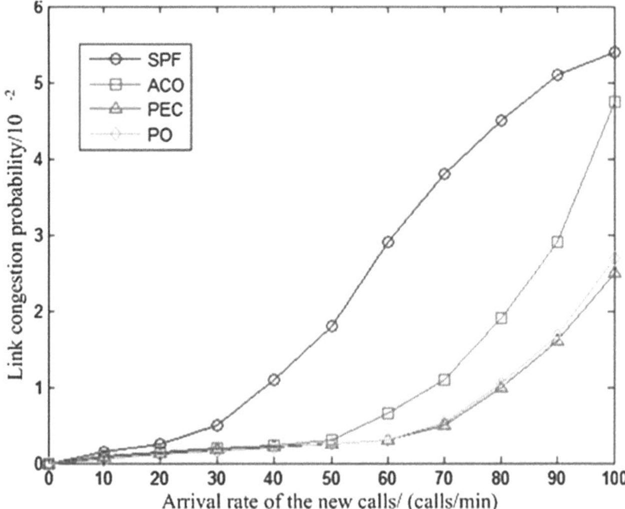

Fig. 5.16 Comparison of the link congestion probability

optimization routing algorithms. PEC represents the Proper Equal Constraint routing algorithm, and PO represents the Prior Order routing algorithm. From the experiment results, we know that the performances of the PEC and PO routing algorithms are quite similar. However, the performance of the ant colony routing

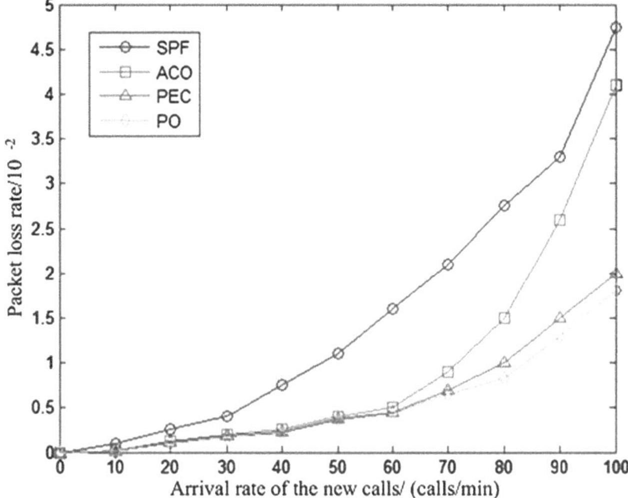

Fig. 5.17 Comparison of the packet loss rate

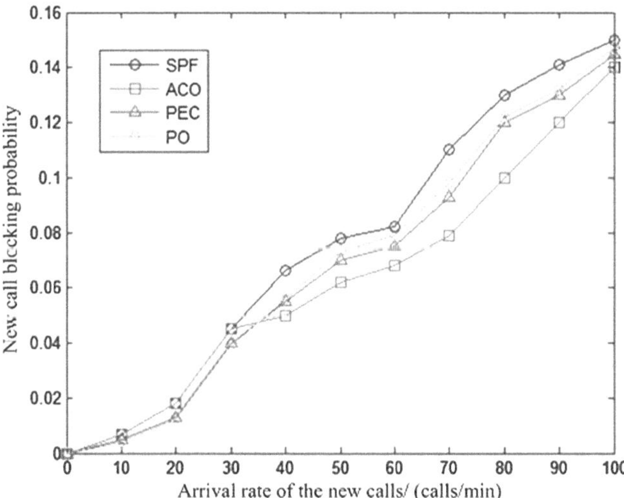

Fig. 5.18 Comparison of the new call blocking probability

algorithm in link congestion and packet loss is weaker than that of the PEC and PO routing algorithms. This is because the calculation procedure of the heuristic algorithms is more complex than the multiobjective routing algorithms. The time overhead of the heuristic algorithms is larger than that of the multiobjective algorithms accordingly, hence the heuristic algorithms are more likely to cause link congestion and packet loss. However, in the aspects of new call blocking and

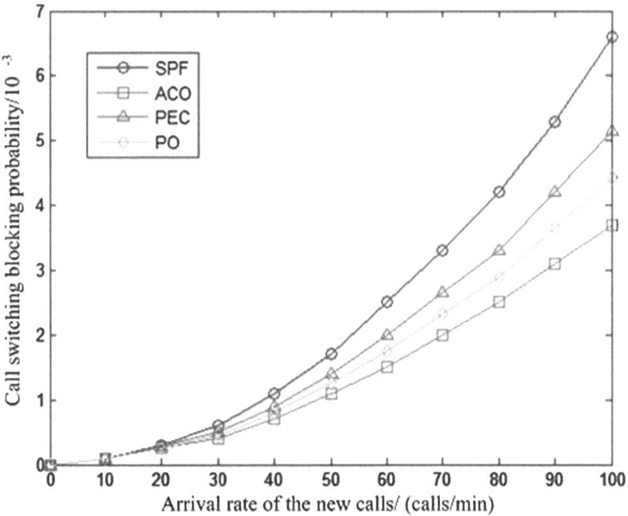

Fig. 5.19 Comparison of the call switching blocking probability

call switching blocking, the performance of the ant colony algorithm is better than that of the multiobjective routing algorithms. This is because the heuristic algorithm chooses the paths in certain probability which can efficiently balance the traffic load, while the multiobjective routing algorithm does not have such a mechanism.

5.5 Summary

Delay, bandwidth and packet loss are selected as the simultaneous optimization objectives in this chapter, and the heuristic algorithms and the multiobjective algorithms are introduced to solve the QoS routing problem of the satellite network. For heuristic algorithms, the ant colony algorithm, the beehive algorithm, the taboo search algorithm, and the genetic algorithm are selected. In the multiobjective algorithm category, the PEC algorithm and the PO algorithm are selected. The performances of these 5 QoS routing algorithms are compared in terms of link congestion, packet loss, and call blocking probability, etc. It is proved that the performances of these five algorithms are better than the SPF algorithm. These five algorithms have their own advantages and disadvantages in terms of certain specific QoS requirements.

References

1. Akyildiz IF, Uzunalioglu H, Bender MD (1999) Handover management in low earth orbit (LEO) satellite network. ACM-Baltzer J Mob Networks Appl (MONET) 4(12):301–310
2. Werner M (1997) A dynamic routing concept for ATM-based satellite personal communication networks. IEEE J Sel Areas Commun 8:1636–1648
3. Ercetin O, Krishnamurthy S, Dao S et al (2002) Provision of guaranteed services in broadband LEO satellite networks. Comput Networks J (Elsevier) 39(5):61–77
4. Chen C (2005) Advanced routing protocol for satellite and space networks [Ph. D. thesis]. Georgia Institute of Technology, Atlanta, U.S.A
5. Chang HS, Kim BW, Lee CG et al (1995) Topology design and routing for low earth orbit satellite networks. Proc IEEE Glob Telecommun Conf (*Globecom*'1995) 11:529–535
6. Chang HS, Kim BW, Lee CG et al (1998) FSA based link assignment and routing in low earth orbit satellite networks. IEEE Trans Veh Technol 8:1037–1048
7. Tsai K, Ma R (1995) Darting: a cost effective routing alternative for large space-based dynamic topology networks. Proc IEEE Mil Commun Conf (*MILCOM*'95) 10:682—687
8. Raines R, Janoso R, Gallagher D et al (1997) Simulation of routing protocols operating in a low earth orbit satellite communication network environment. Proc IEEE Mil Commun Conf (*MILCOM*'97) 11:429–433
9. Cheng C, Riley R, Kumar S et al (1989) A loop-free bellman-ford routing protocol without bouncing effect. Proceedings of ACM SIGCOMM, New Orleans LA, pp 224–236
10. Ekici E, Akyildiz IF, Micheal DB (2000) Datagram routing algorithm for LEO satellite networks. In: Proceedings of IEEE INFOCOM'2000. Tel-Aviv, Israel, pp 500–508
11. Ekici E, Akyildiz IF, Bender MD (2001) A distributed routing algorithm for datagram traffic in LEO satellite networks. IEEE/ACM Trans Networking 2:137–147
12. Chan TH, Yeo BS, Turner LF (2003) A localized routing scheme for LEO satellite networks. In: Proceedings of AIAA 21st international communication satellite systems conference and exhibit (*ICSSC*'2003), Yokohama, Japan, pp 2357–2364
13. Chen C, Ekici E, Akyildiz IF (2002) Satellite grouping and routing protocol for LEO/MEO satellite IP networks. In: Proceedings of the 5th ACM international workshop on wireless mobile multimedia (*WoWMoM*'2002), Rome, Italy, vol 9, 109–116
14. Berndl G, Werner M, Edmaier B (1997) Performance of optimized routing in LEO intersatellite link networks. Proceedings of IEEE 47th vehicular technology conference (*VTC*'97), Phoenix, Arizona, vol 5, pp 246–250
15. Chen C, Ekici E (2005) A routing protocol for hierarchical LEO/MEO satellite IP networks. ACM Wirel Netw J (WINET) 11(4):507–521
16. Xing WX, Xie JX (2007) Modern methods of optimization calculation. Tsinghua University Press, Beijing [in Chinese]
17. Reeves CR (ed) (1993) Modern heuristic techniques for combinatorial problems. Blackwell Scientific Publications, Oxford
18. Li SY (2004) Ant colony algorithm and its applications. Harbin Institute of Technology Press, Harbin [in Chinese]
19. Gui ZB, Ji XQ (2003) A unicast QoS routing algorithm based on ant colony system. Sig Proc [in Chinese] 19(5):431–433
20. Admeia J, Hoang N (2000) A QoS based routing method for multi-hop LEO satellite networks. In: Proceedings of IEEE international conference on networks. Singapore, pp 399–403
21. Glover F (1986) Future paths of integer programming and links to artificial intelligence. Comput Oper Res 13:533–549
22. Glover F (1989) Taboo search: part I. ORSA J Comp 1:190–206
23. Wedde HF, Farooq M, Zhang Y (2004) BeeHive: an efficient fault-tolerant routing algorithm inspired by honey bee behavior. Ant Colony Optimization Swarm Intell Lect Notes Comput Sci 3172:83–94

24. Wedde HF, Farooq M (2005) BeeHive: routing algorithms inspired by honey bee behavior. Künstliche Intelligenz Spec Issue Swarm Intell 4:18–24
25. Lin C, Dong D (1992) Methods and theories of multi-objective optimization. Jilin Education Press, Jilin [in Chinese]
26. Narvaez P, Clerget A, Dabbous W (1998) Internet routing over LEO satellite constellations. In: Proceeding of 3rd ACM/IEEE international workshop on satellite-based information services (*WOSBIS'98*), Dallas, Texas, pp 89–95
27. Wood L, Clerget A, Andrikopoulos I et al (2001) IP routing issues in satellite constellation networks. Int J Satell Commun 19(2):69–92
28. Yitas D, Halim ZA (2007) A dynamic routing alrotithm in LEO satellite systems estimating call blocking probabilities. Proceedings of the 3rd international conference on recent advances in space technologies (*RAST'07*), Istanbul, Turkey, vol 6, pp 181–189

Chapter 6
Summary and Outlook

6.1 Summary

With the development of science and technology, satellite networking has an increasingly wide range of applications in the fields of telecommunications, meteorology, scientific explorations, and military services. Satellite network routing technology is the core of the satellite network technology, and determines the performance of the whole system. The inherent characteristics of a satellite network, such as dynamic topology, inconvenient equipment maintenance, and constrained onboard resources, make the existing terrestrial network routing technology unsuitable to be applied to satellite networks. In order to achieve the performance similar to the terrestrial network, it is necessary to design a set of robust QoS routing methods for satellite networks according to their characteristics. In view of this, we research constellation designs, routing strategies, traffic engineering and routing algorithms, and propose a set of routing technologies that is robust to the traffic load and can meet the QoS requirements of different users. The main contribution of this book is as follows:

(1) The goals of satellite constellation design are proposed to make it easier to implement the robust QoS routing technology. The effects of layers of satellites, types of orbits, orbital altitude, the number of satellites, and the arrangement of satellites on the constellation performance are analyzed theoretically. A triple-layered satellite constellation is designed according to the design goals. Some basic indicators of onboard routing and their ability to deal with some of the basic business requirements are given through simulations.
(2) A novel satellite network routing strategy-NSGRP is proposed. NSGRP takes the advantage of easy implementation of the virtual node strategy and the advantage of high extensibility of the virtual topology strategy. The LEO satellites are switched collectively during the free period in NSGRP, which greatly reduces the snapshots in a system cycle, hence reducing the protocol overhead. NSGRP improves the link congestion handling mechanism and the satellite node failure handling mechanism of the virtual topology grouping

F. Long, *Satellite Network Robust QoS-aware Routing*,
DOI: 10.1007/978-3-642-54353-1_6,

strategy, which makes the routing strategy more robust. Finally, the simulation results show that NSGRP performs better than the previous routing strategies.

(3) A satellite network traffic engineering scheme, which is robust under the traffic spike, is proposed. First, a satellite network traffic prediction method is proposed based on time series analysis according to the global distribution of previous satellite network user traffic. Second, the predicted traffic matrix is optimized, and then the cost envelope is added to limit the performance when a traffic spike is coming. The robust traffic engineering scheme under unpredicted traffic can thus be got. The robust traffic engineering approach performs near-optimally under normal traffic, and its performance will not decline sharply when the unpredicted traffic spike is coming.

(4) A multi-QoS-objective routing optimization method for satellite networks is proposed. Some heuristic algorithms and multi-objective optimization methods are introduced to solve the multi-QoS-objective routing optimization problem of a satellite network, which can mitigate the disadvantage of previous single-QoS-objective routing optimization methods, which optimize one QoS indicator at the expense of another QoS indicator. Bandwidth, packet loss, and delay are used as our optimization objectives in this book. The simulation results show that the heuristic and multi-objective optimization algorithms have the ability to optimize these objectives simultaneously, and perform better than the SPF routing algorithm.

In summary, a complete set of technological schemes, including constellation design, routing strategies, traffic engineering, and routing algorithms for a satellite network, is proposed according to the characteristics of the satellite network. The technological schemes have the advantages of low traffic overhead and can meet several QoS requirements simultaneously. They can basically solve the robust QoS routing problem mentioned in Chap. 1.

6.2 Outlook of Further Endeavors

In the research in this book, the satellite network, as an independent autonomous system (AS), fulfills business requirements from the source terrestrial users to the destination terrestrial users. With the development of satellite network technology, the mergence of and cooperation between satellite networks and terrestrial networks will become a trend. Satellite networks have the advantages of easy access and wide coverage areas, while terrestrial networks have the advantages of high bandwidth and reliable transmission. The cooperation between the two can complement each other to provide the users with a more convenient and reliable service. How to design a secure and self-adaptive border gateway after mergence to achieve a seamless connection of the satellite networks and terrestrial networks poses a very challenging task to be accomplished urgently.

Satellite networking is also an important part of the Inter-Planet Network (IPN). With the development of space technology, the exploration of deep space will be one of the frontier research topics. The problem of interplanetary network routing will also be an important problem to be solved in interplanetary communication in the near future. Interplanetary networks face the same problems of intermittent connectivity and constrained resources as satellite networks, and it is more inconvenient to repair and replace their equipment compared with satellite networks. The interplanetary networks must maintain network connectivity under an extreme external environment, which requires the routing protocol more self-adaptive. However, the topology of interplanetary networks is also as predictable as satellite networks, which means that the routing protocol of satellite networks can be applied for reference to interplanetary networks. An interplanetary network can be divided into several subnets in accordance with mission requirements or for the reason of environmental interference, which requires the components of the network to have the ability of self-organization. This is similar to ad-hoc networks. It is of great significance to develop feasible routing protocols for interplanetary networking based on the existing network protocols of the ad-hoc networks and satellite networks.

Besides, a satellite network is also a kind of Delay and Disruption Tolerant Network (DTN). The instance of DTN includes deep-sea communication networks, battlefield communication networks, and interplanetary networks. In general, challenges to be faced by the DTN routing protocols are: large delay caused by physical link characteristics, network separations and a high link error rate caused by crucial severe environments. Since there are many instances of DTN and the application range of DTN is quite wide, it is also of great importance to develop DTN routing protocols using satellite network routing methods for reference.

Zeitfracht Medien GmbH
Ferdinand-Jühlke-Straße 7
99095 Erfurt, Deutschland
produktsicherheit@kolibri360.de